The Global Debt Crisis

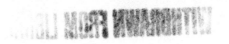

THE GLOBAL DEBT CRISIS

America's Growing Involvement

JOHN H. MAKIN

Basic Books, Inc., Publishers   New York

Library of Congress Cataloging in Publication Data

Makin, John H.
    The global debt crisis.

    Includes index.
    1. Loans, Foreign—Developing countries. 2. Debts, External—Developing countries. 3. Loans, American— Developing countries. I. Title.
    HG3891.5.M34 1984   336.3'435'091724   84–45075
    ISBN 0–465–02681–8

To the memory of

my mother and father

Let not thine hand be stretched out to receive
and shut when thou shouldest repay.

—Apocrypha   4:30–31

Contents

Contents

Acknowledgments

THIS BOOK began to take shape during the three years between 1965 and 1968 when, in the intense intellectual atmosphere at the University of Chicago, I began to understand the nature of national and international financial systems, thanks especially to Milton Friedman, Robert Mundell, and the late Sir Harry Johnson. Since then many others have nurtured and enhanced my interest and, hopefully, my understanding: during my first summer at the International Monetary Fund (IMF), Victor Argy, Zoran Hodjera, and the late Bill White taught me that Chicago training is not everything; my stint at Treasury during 1971–72 taught me that economics is not everything, but also made me determined to make it play a larger role in thinking and policy making on international economic issues; a year at the University of British Columbia (UBC) introduced me to life in a small open economy, an experience I recommend to all Americans, especially those who set economic policy; UBC also spawned a fruitful collaboration with my friend Maurice Levi, who later introduced me to Martin Kessler and Basic Books.

We all have at least one book in us. It requires a catalyst to bring it out. Mine was a combination of elements: another visit to the IMF between September 1982 and May 1983, thanks to Vito Tanzi; discussions with Alan Tait and Don Mathiesson at the IMF, Nick Sargen at Morgan's, Jacob Dreyer at Treasury, John Anderson at the *Washington Post,* and the late Bob Weintraub at the Joint Economic Committee all helped to keep me convinced that this book needed writing and did it without compromising the delicate positions of each of their institutions in an evolving crisis. During 1983 lectures on various aspects of global debt were given at the European Institute of Business Administration in Fontainebleau; the Institute for Economic Research on Monetary Issues in Ljubljana; the Political Science Faculty in Belgrade; the Republican Secretariat for Finance and Croatian Association of Banks in Zagreb; the Faculty of Economics at the University of Rome; Liguria Federation of Industries in Genoa; Institut für Wirtschaft und Politik (Haus Rissen) in Hamburg; the Department of Monetary Economy at Erasmus University, Rotterdam; the Institute for Financial Affairs in Tokyo; and the Group for Economic and Financial Reflection in Paris.

Acknowledgments

Authors usually begin a book with an idea that needs to be placed somewhere in the broad sweep of history and written about in a lively and compelling manner. For adding these crucial ingredients and for the extra push to "get moving" that only his remarkable energy could provide, I owe thanks to my friend and tutor, Timothy Dickinson.

The manuscript was flawlessly, promptly . . . and patiently . . . typed through numerous drafts by Marian Bolan at the Institute for Economic Research of the University of Washington's Department of Economics.

It is ironic, but almost always true, that the time and energy to write a book in the midst of one's busy and often hectic midlife must be stolen from those one loves best of all. My wife, Gwendolyn, has, in addition to patiently allowing me to disappear for hours of writing, chased down elusive reference materials, copied, typed, taken notes on exhausting train rides through Europe, translated, and even generously allowed me to steal an occasional idea. Needless to say, it would not have been possible without her help.

This catalogue of debts is longer than I had supposed before sitting down to write it out, and I have left out many good friends whose encouragement and good humor sustained me. On reflection, however, I am convinced that its length should in no way diminish my obligation to everyone in it.

JOHN H. MAKIN

University of Washington
Seattle, Washington
May 1984

The Global Debt Crisis

"I am sick of being looked at like one of your Brazilian bank loans!"

Drawing by Wm. Hamilton; © 1984 The New Yorker Magazine, Inc.

Introduction

THIS is a book about the most remarkable investment boom in the eight-century history of international banking. The investors were a small group of the world's largest banks, which, along with a few governments and international agencies, put up almost $700 billion during the decade between the early 1970s and the early 1980s. In ten years these investors lent to a group of politically and often economically unstable developing countries a sum equal to the total amount borrowed by the United States government between the founding of the republic in 1776 and the accession of Carter in 1977. Those two centuries included the War of Independence, the Civil War, two world wars, and the Korean and Vietnam wars, which by 1967 accounted for a national debt of just about $325 billion. Even the $375 billion increase in the national debt during the 1967–77 decade, produced by the relentless demands of war and greatly expanded social programs, was just over half of the burst of lending to less developed countries (LDCs) undertaken during the decade centering on 1977.

The result of this lending surge was billions in new debts of foreigners to Americans instead of the billions in new debts of Americans to themselves that come with annual growth of the national debt. In all of the industrial countries, precious capital was flowing out into what was becoming an economic desert instead of nourishing growth of commerce at home.

The explosion of bank lending to LDCs, which by 1982 had resulted in the world debt crisis, utterly transformed the conception of what might be counted as bank assets. In the past, lenders had acquired claims against a wide variety of businesses collateralized by tangible assets that could be attached in the event of a failure to repay. Bank loans to their own governments had also grown rapidly since World War II, but such financing of national debts was backed by the full faith and credit of well-established states. In contrast, what the banks acquired at an unprecedented rate during the immense distension of lending to LDCs were claims offset against nothing more tangible than a set of political attitudes and assumptions in a group of countries that were at once economically and politically volatile and not infrequently hostile toward the capitalism exemplified by their creditors. Mexico, for example, which still felt strong resentment over what it saw as the theft by the United States of over half of its territory—Texas—over a century earlier, might not be expected to feel duty-bound to honor its obligations to the big American banks.

In order to reach a state that might fairly be called a world debt crisis by the end of August 1982, the banks had to expand their loans to developing countries at a rate of almost 25 percent per year through the cartelization of the world's oil, revolution in Iran, war in the Gulf, and the unprecedented turmoil that accompanied the economic hurricanes of the late 1970s and early 1980s. The new breed of bankers who engineered this revolution in banking were drawn by the sense of a "new economic order" and by some powerful economies of scale inherent in lending hundreds of millions of dollars to a single dictator or minister at a stroke instead of plodding along placing a hundred thousand here and a hundred thousand there to put up shopping centers or industrial parks. They saw themselves as getting in on the ground floor of Eldorado, and forgot that the winners are paid off on the top floor and all but the strongest entries fall on the stairs. The growth of the automobile and computer industries has been a sifting and winnowing process that left only a few strong survivors to reap the huge rewards. Likewise the rewards will undoubtedly be huge for those of the world's developing economies that achieve sustained development; but the banks were treating almost all of them like prospective winners.

Besides affording tremendous immediate profits, participation in the emergence of the new economic order brought new glamour into the sometimes humdrum lives of the big banks' loan officers. As any traveler knows, the quality of treatment received by official visitors varies inversely with the income per capita of the country visited. Finance ministers of developing states have something to prove to their visitors from Chase and Morgan in New York and Barclays or Lloyds in London. Their private dining rooms, offices, and limousines, the sumptuous treatment offered to their guests, are a far cry from the reception that awaits a banker who visits Oklahoma City to look into oil leasing or Butte to investigate a mining interest. Once you have been met and guided through customs by a solicitous driver who then whisks you through traffic in an air-conditioned limousine to your luxurious hotel and thence to an elegant dinner with the truly charming people who head the finance ministries of developing countries, it is a real come-down to grab a cab and check yourself into one of the inevitably tacky hostelries that pass for the "best" in towns where so many American companies see fit to locate themselves. Dinner conversation in places like that, if someone meets you for dinner at all, is less than exotic.

By 1979 bankers were literally chasing prime ministers and finance ministers around hotel lobbies in a desperate effort to outlend their rivals. British jour-

Introduction

nalist Anthony Sampson describes the game as played at the September 1980 meeting of the International Monetary Fund (IMF) and World Bank at the Sheraton Washington Hotel:

> Back in the hotels it becomes easier to distinguish the different tribes of the hunters and hunted, for they wear colored badges stating their business or country. Blue for the governors (the ministers of finance or their deputies); green for the delegates from 140 countries; brown for the guests, who include the top bankers; orange for the 600 journalists; and red for the "visitors," the hordes of ordinary bankers in pursuit of their customers. As they pair off or make huddles it is tempting to play a kind of card game with them, awarding points to different combinations. Greens with blues are common enough; oranges and greens are not surprising. It is the combination of red and blue—a banker who has buttonholed a Minister of Finance—which scores the most points, for here a loan may be clinched, or a deal signed up. The ministers from credit-worthy countries are desperately in demand. "The trouble with this business," as one investment banker explained, "is that there are ten hares chasing each fox. When you get to the minister, your rivals have got there before you." The new Prime Minister of Peru, Dr. Ulloa, is in special demand: five years ago his country seemed almost broke, but now the oil is flowing westward across the Andes pipeline, transforming his balance of payments. "I can hardly face going back to the hotel," he is heard to complain; "there are six different banks waiting for me."[1]

But lavish dinners and VIP treatment will not explain the bankers' almost frantic pursuit of Manuel Ulloa and his colleagues. If that kind of treatment alone could extract the hundreds of billions of dollars the bankers committed to the LDCs between 1974 and 1982, they would surely find themselves as pampered in Rockville as they were in Rio. Something else was on the mind of financiers, something that had been there for almost a decade. It was a deep-seated feeling of unease that somehow all was not well in their own countries, that they were running out of the things they would need to keep the machinery running and furnishing the life-style they had come to expect. They wanted a piece of the part of the world that was deemed still to hold tremendous promise for the future, rich in the raw resources that could be fed into the machinery. Brazil held that promise. Its growth had averaged 6 percent between 1920 and 1967. The "Brazilian miracle" of 1968 to 1973 saw growth pound along at 11 percent a year—more than twice the average rate in industrial countries and closer to three times the rate in many, including the United States. Other countries besides Brazil held similar promise, especially after the first oil shock jolted even the less imaginative bankers into the realization that the era of plentiful raw materials at low cost was over. The quest was for "energy," both literally and figuratively, and it was sought with an intensity akin to that of the aging seeking the fountain of youth.

The story of the LDC lending boom must be told on two levels. The first describes the huge expansion of direct lending by banks to the governments of developing countries. But fascinating as that story is, it is only a symptom of something more fundamental, more compelling: a fear that the industrial countries' economic machine was running down. During the 1970s and 1980s, the world's economically advanced countries came to feel the cold breath of economic entropy, a sense that the hopes of prior decades would not be realized. One painful reminder of this loss of economic drive was the dramatic fall in productivity and capital formation in most industrial countries during this time. Another was the explosion of government debt in the advanced countries; during the 1970s and 1980s it has paralleled that of the developing countries.

In the United States, the economic hopes of the 1960s were legislated into economic requirements with the creation of the "Great Society" programs in the wake of even more ambitious social programs in Europe. Japan followed suit in the mid-1970s. But the economic hopes embodied in these ambitious programs were not realized during the 1970s. The result was a ballooning of government debt in industrial countries as revenues fell short of the levels required to provide the promised benefits. The initial response was the usual one of governments when spending exceeds revenues: tax increases were avoided and more money was printed. Inflation had been accelerating in the industrial world even before the first oil crisis, rising from an average 3 percent a year during the 1960s, to 7 percent by 1973. But after 1974 governments chose to validate the steepening rise in energy costs with a monetary expansion that further raised prices. The higher prices considerably reduced the burden of government debts that were conveniently expressed in terms of money whose value was steadily reduced by higher prices. The debts began to grow rapidly, bridging the gap between the sluggish seventies and the rapid economic growth that the ebullient sixties had assumed would come as part of the natural order of things to pay painlessly for the vast new programs.

Just as the developed countries were beginning to realize that in the absence of sustained rapid growth social programs meant larger debts and those larger debts meant higher taxes—unthinkable—or higher inflation, they were hit with the first oil shock. Attention turned to the Organization of Petroleum Exporting Countries (OPEC), which almost overnight could in 1974 levy and collect a tax of $70 billion—equal to the total amount saved by Americans during that year. The huge transfer of resources from oil importers to oil exporters was a banker's dream. Overnight it created a whole new category of borrowers and lenders: the oil buyer and the oil seller. The sellers were not very sophisticated in financial matters—they wanted to put their money in the bank. At first the banks' biggest problem was to recycle the money fast enough —to get it into the hands of borrowers in huge chunks while collecting their

Introduction

fees, which shot up simply as a reflection of the delightful fact that 1 percent of $100 million is ten times larger than 1 percent of $10 million. Jumbo loans are just as easy to make as large loans and yet much more profitable.

After a year or so the oil exporters learned to spend their bonanza at a much faster clip. Their combined current account surplus fell from $68.3 billion in 1974 to $35.4 billion in 1975. But the lending continued to grow, as the banks began to use funds drawn from investors in developed countries who were looking to own a piece of Eldorado. The desire was to buy some of the economic vitality imparted by possession of rich stores of all kinds of natural resources—oil, copper, tin, manganese—needed to run the sophisticated economic machinery in the developed countries. The lending boom was on.

The two-level story of the lending boom that ended in crisis is told in four parts. The first part, chapters 1 and 2, describes the breaking of the bubble in August 1982 and tells something of the shocks experienced in the past by lenders to governments. It introduces the current crisis and places it within the context of previous lending booms and defaults. The second part, chapters 3 and 4, looks at industrial countries' fears about shortages of natural resources and developing countries' fears about losing benefits of natural endowments. It explains the basis for the powerful attraction between lending banks and borrowing countries in the wake of the first oil shock. After the shock the two regions moved in opposite directions, with the banks acting as a kind of cable taking the strain before the recession of 1981 began to make it fray.

The third part, chapters 5, 6, and 7, describes the origins of the crisis. American ambivalence about assuming hegemony in the postwar international financial system is traced back to attitudes of Franklin Delano Roosevelt at the time of the prewar international financial crisis of the early 1930s. The London "gold bubble" that emerged in October 1960 at the height of the Kennedy-Nixon battle for the American presidency signaled the start of a decade of American ad-hocery in international economic policy that created, among other things, the stateless, free-floating eurodollar market while gearing up American banks to become direct lenders to foreign governments for the first time in history. This, together with pre-oil inflationary excesses of the early 1970s in the industrial world, which were made necessary by overcommitments of governments on social programs coupled with the incessant drain of the Vietnam War on the United States—Nixon abandoned gold in 1971 to free himself to pay for Vietnam with an inflation tax unacceptable to large dollar holders abroad—set the stage for the debt crisis.

The inflation of the early 1970s emboldened OPEC members to charge more for oil and to discover that "more" exceeded their wildest hopes. The resultant emergence overnight of lenders and borrowers together with the banks' practice in and profitability with direct lending to foreign governments—especially

7

those in developing countries—started the lending boom that was to reach $700 billion by the end of 1982. The process was sustained by a desperate effort —especially by the Carter administration in the United States—to revive the economy from oil shock and attendant reduced ability to produce with heavy doses of demand. The "locomotive theory" that saw the United States, with expansionary monetary and fiscal policies, pulling the rest of the world out of its post-oil-shock doldrums, was like a belief that the only thing needed to make a GM or Ford or Chrysler plant, geared in 1975 to make two-ton cars that got 10 miles per gallon, hum right back to life was a good healthy shot of easy money. What they needed was retooling to make lighter, fuel-efficient cars, and that was not likely, given policies like Carter's cap on gasoline prices.

Throughout the ever-more-inflationary seventies, the banks rode a lending boom while the borrowing countries in the Third World—seeing prices of their exports shoot up in the inflationary environment—were only sorry they had not borrowed more. OPEC gave the borrowers and their eager bankers another excuse to prolong the debt boom by responding to the inflationary 1970s with another doubling of oil prices at the end of the decade. The banks thought they were protecting themselves with floating-rate loans wherein interest rates were tied—at a nice premium—to the London interbank rate, which would rise with the inflation that devalued fixed-rate instruments like Treasury bonds. They did not, however, allow for a 1981–83 credit crunch that pushed up inflation-adjusted interest rates while ushering in a collapse in prices of commodities whose reduced sales could no longer even come close to meeting LDC debt servicing needs.

The fourth part, chapters 8 and 9, brings onstage some important players —lenders of last resort, or so some hope—in the debt crisis, the International Monetary Fund and the Bank for International Settlements. The evolution of these institutions is entwined with the discussion in part 3 of crisis origins to help explain the roles they play after the August 1982 emergence of the debt crisis.

The fifth part, chapters 10, 11, and 12, describes the onset and unfolding of the crisis together with its immediate antecedents. A perspective on the world debt crisis of the eighties together with discussion of some possible outcomes —there are no simple solutions to make it go away—completes the book.

PART I

An Unprecedented Crisis

Chapter 1

Dangerous Loans to Princes

Rather than refuse deposits the Medicis succumbed to the temptation of seeking an outlet for surplus cash in making dangerous loans to princes.

RAYMOND DE ROOVER,
The Medici Bank

ON Friday the thirteenth of August, 1982, the finance minister of Mexico, Jesus Silva-Herzog, paid a quiet visit to the United States Treasury. The day was sunny and cool, only about 80 degrees; more like an autumn day along the Potomac instead of the usual, wretched weather that Washingtonians must endure in August. The unusual day was a fitting background for Mr. Silva-Herzog's mission. The proud prince of finance of a proud nation with over $15 billion in oil exports had come hat in hand to Washington to beg for money.

Mexico Reaches a Crisis

The task must have been even more distasteful in light of the longstanding strained relations between Mexico and the United States. Westward expansion of the United States had included annexation in 1848 of what previously had constituted nearly half the land area of Mexico. More recent irritants included Jimmy Carter's gaffes about "Montezuma's revenge" during his visit to Mexico City and the gelding of a prime stallion given to President Ronald Reagan by President José Lopez Portillo. Earlier in the summer of 1982 Mexico's by-then lame duck President Portillo became furious when he laid his hands on a June 26 State Department briefing paper on Mexico that concluded with the suggestion that Mexico's economic crisis might lead it "to sell more oil and gas to us at better prices." A Mexico "with the wind out of its sails," the State Department memo speculated, might be more ready to ease restrictions on foreign investment, negotiate a trade agreement, cooperate in control of illegal migration, and in general be "less adventuresome in its foreign policy and less critical of ours."[1]

The briefing paper, together with remarks by John Gavin, a former actor and the U.S. Ambassador to Mexico, on a July 25 ABC-TV special suggesting that Central America's troubles could "spill over" into Mexico (remarks that so angered the Mexican government that consideration was given to declaring Mr. Gavin persona non grata), created what the *New York Times* described in its August 14 edition as "the first serious rift in United States–Mexican relations since President Reagan took office."[2]

The timing of Mexico's eruption of indignation to coincide with Mr. Silva-Herzog's arrival in Washington, by providing him with a grievance in need of redress, perhaps made his task a little less distasteful. In fact, depiction of Mexico as a nation with the "wind out of its sails" was far more appropriate than the authors of the briefing paper could have imagined. Mr. Silva-Herzog's immediate mission was to seek credits of $3.0 billion from the United States to enable payment of interest coming due during August and September on Mexico's external debt. In spite of its massive oil revenues, Mexico had reached a point where its major creditors, the commercial banks, would no longer advance the money needed to pay the interest on its debts. Failure to make such payments over the next few months would require banks to declare in default some $80 billion worth of loans to Mexico. A rupture of that magnitude in the world's intricate financial network would be devastating for banks as well as for thousands of unsuspecting investors who were in mid-August relaxing in Kennebunkport or hiking in the Sierras. A Mexican default would mean losses of nearly $12.0 billion for the six largest banks in the United States, cutting the value of their shares by more than one-half. Nor would the damage be confined to the big banks and their shareholders. More than 1400 American banks had investments in Mexico, with some of them heavily exposed. Beyond that, there were heavy investments in other countries, especially in Latin America—exposure in Brazil was even larger than it was in Mexico, and heavy exposure also existed in Chile, Argentina, Venezuela, and Peru, not to mention Poland, the Philippines, Yugoslavia, and numerous others.

The revelations of the magnitude of less-developed-country (LDC) borrowing that emerged after a flurry of investigations prompted by Mr. Silva-Herzog's visit to Washington were staggering. At the moment of his arrival he himself did not know the full extent of his country's debt, let alone that of other heavy borrowers—already an indication that very important details had begun to be overlooked. Neither did the U.S. Treasury or the International Monetary Fund (IMF), the world's major supplier of short-term adjustment financing and the leading candidate on everyone's wish list for international lenders of last resort who, it was hoped, would emerge to shore up the tottering world financial system. Only a few insiders at the Federal Reserve Bank of New York, another potential major world safety-net manager, had any idea that a serious problem existed. They had hinted at the problem during a July gathering of bankers in Aspen, Colorado. Between August 13 and the start of the IMF World Bank joint annual meetings on September 6, estimates emerged that debts to all LDCs combined totaled between $600 and $700 billion.

Early Rescue Efforts

Although the weekend's newspapers carried no account of Mr. Silva-Herzog's Friday the thirteenth visit to the Treasury, its results were newsworthy enough. The $3 billion requested was made available in time to meet Mexico's immediate commitments. Mr. Silva-Herzog's appearance in Washington confirmed a fear that had been lurking in the minds of top-level Treasury and Federal Reserve officials for months, ever since the much-heralded world economic recovery of 1982 had failed to materialize. That fear combined with the ice-water shock of Mr. Silva-Herzog's visit to produce extraordinary action. The $3 billion package included $1 billion from the Commodity Credit Corporation, a $1 billion advance payment by the Department of Energy for Mexican oil to go into the United States' strategic reserve, $700 million directly from the Federal Reserve, and $300 million from the Treasury under existing, prearranged ("swap") lines of credit that had been set up between central banks. In addition to the $3 billion directly from the United States, Mexico sought and later received a $1.85 billion line of credit from the Bank of International Settlements (BIS) in Basel. The BIS is the bank of central banks and institutions, tapped by some as another international lender of last resort, and its prompt, sizable involvement indicated central bankers' real concern that the problems of Mexico and other large borrowers menaced financial stability far outside their borders.

Discussions completed at the Treasury, Mr. Silva-Herzog's limousine traveled just four blocks west along Pennsylvania Avenue, negotiated a difficult left-hand turn onto Nineteenth Street, and turned right into the cobblestone driveway of the International Monetary Fund. The headquarters of "the Fund," as it is called by its cosmopolitan, well-paid staff drawn from powerful families all over the developing world (the staff from rich member countries is composed more of professionals not necessarily possessed of political influence), is a magnificent building. As such it and the people who work there are objects of considerable resentment by U.S. government workers, who are paid considerably less than Fund staff and who work in far less splendid quarters. The building itself, completed in 1973 to replace already impressive quarters across the street, has since been the subject of reverent articles in magazines read by architects and interior designers who no doubt salivate at the thought of fees earned by its architects, Vincent G. Kling and partners. It is a structure modern and linear on the outside with an arrogantly nonfunctional interior. With a stroke amounting almost to an eighteenth-century folly, its architects chose to use half a block of choice real estate in the world's most powerful capital city to create an empty middle, a sumptuous interior atrium that allows the sun's rays to penetrate thirteen stories and shine on those who come into the Fund—a "brilliant example of inner space," declared Wolf Von Eckardt in a burst of enthusiastic praise that appeared in the *Washington Post* on August 18.[3] The tall, almost white structure conveys to the most casual observer a sense of solidity and capacity, reminding visitors of an earlier epoch when the United States was the steel anchor in bad situations, impassive, at once imparting confidence and concealing, indeed denying, any grounds for alarm.

The offices of the IMF's managing director, Jacques de Larosière, are on the twelfth floor facing west out over George Washington University, the Kennedy Center, Georgetown, and the Potomac. Mr. Silva-Herzog crossed the sunlit atrium to the cool marble of the low-ceilinged corridors leading deeper inside the Fund—as if leaving the heart of an Egyptian temple, suggested Mr. Von Eckardt—and rode one of the wood-and-leather-paneled elevators to the twelfth floor where he stepped out into the richness of teak paneling and soft carpeting that creates a serene and comfortable atmosphere to contain the offices of the IMF's most powerful men. Here too his mission was not one he relished. In spite of Mexico's consistent bragging about not involving the IMF in its affairs, Mr. Silva-Herzog was going to ask Mr. de

Dangerous Loans to Princes

Larosière to initiate procedures for granting Mexico an extended credit facility of $3.92 billion.

Mr. de Larosière responded promptly. It normally requires months even to arrange the fact-finding visit required to initiate a loan—what the IMF somewhat messianically calls a mission—but four days later one was flying to Mexico City; twelve weeks of intensive negotiation followed, and a draft letter of intent indicating Mexico's willingness to meet IMF conditions was countersigned by Mr. de Larosière in Washington on the evening of November 9.

The IMF mission had established that Mexico's debts totaled $81 billion, of which some $67.5 billion were held by commercial banks. According to figures that were later to appear in the *American Banker,* house organization of the American banking fraternity, and *Newsweek,* Citicorp was the largest of Mexico's American creditors, with $3.3 billion in holdings, accounting for more than two-thirds of its net corporate assets. (Citicorp had another $6.5 billion at risk in Brazil, Venezuela, Argentina, and Chile, bringing its exposure to over 200 percent of its net corporate assets. If only half of these investments went bad, Citicorp's shareholders would lose all of their investments.)[4] Of the $67.5 billion Mexico owed, about $28 billion was due within one year, enough to eat up almost all of the proceeds from the country's $33 billion a year in exports.

By themselves the IMF's available resources were not remotely sufficient to meet Mexico's requirements. At a hastily called November 16 meeting of Mexico's major private bank creditors inside the fortresslike Federal Reserve Bank of New York, Mr. de Larosière broke with all tradition by warning that the Fund would lend nothing if the private banks did not come up with $6.5 billion, $5 billion of it new money, by December 15. Bankers like Peter Smith of Morgan Guaranty and William McDonough of First Chicago were surprised at the boldness of the move. The large banks did not like it, but they had little choice but to go along since the alternative was default and a crippling write-off of bank assets. What concerned them was whether the smaller of the 1,400 or so of Mexico's creditor banks would continue to participate. These banks—like small investors, the last to join in a boom and the first to want out at a hint of trouble—would probably seek to pull out of international lending altogether. Every dollar so removed meant another dollar that would have to be found by the large banks, over and above their already huge commitment.

By the December 15 deadline about $4.6 billion of the $5 billion requested was on hand; the missing $400 million reflected the refusal of the smaller banks to go along.

Slow Recognition of a Global Crisis

The hesitancy of some banks only slightly exposed in the international market to keep rolling over their usual short-term loans was in fact justified by a series of rapid-fire revelations during the autumn of 1982 and winter of 1983. *Business Week* featured an article entitled "Worry at the World's Banks" in its September 6 issue. "Lenders' Jitters" was the *Wall Street Journal*'s front page headline on September 15. "Debtors' Prism," quipped the London Economist in an article in its September 11 issue detailing problems in Mexico and Latin America. By October 16, the *Economist* had stepped up the rhetoric in an article entitled "The Crash of 198?" which discussed the Mexican crisis as one of a series including Drysdale, Ambrosiano, and Penn Square. Specific figures detailing the sum and distribution of developing countries' longer term debts emerged with amazing speed. Frightening figures on a rapid run-up of short-term debts were to come later, although no timely and systematic monitoring of such debts, wrongly thought to be self-liquidating, had been kept prior to emergence of the crisis.

At first, neither these dramatic revelations nor the uncustomary IMF initiative with the banks captured the attention of the Reagan administration or Congress. This unfortunate state of affairs is typical of a long history—dating back to the Great Depression and before—of the inadequate attention paid by U.S. governments to problems of the international economy. It will reappear many times in our story of the unfolding of the debt crisis of the eighties. During the autumn of 1982, preoccupied with the recession and holding to the parochial view of world problems most typical of the early years of new administrations, the American government declined to show much enthusiasm for the proposed increase in IMF lending resources at the Toronto meetings in September. (The United States holds an effective veto power over such increases.) Not until late November did the Treasury begin to turn more sympathetic attention to the problem. By mid-December, when Treasury Secretary Donald Regan was testifying before Congress on what had by then become for the administration an "essential" increase in IMF resources, it was discovered that the debt issue had become a hot potato—discussed heatedly from a largely domestic perspective—in the arena of U.S. politics. In response to his suggestion that an $8.0 billion appropriation for the IMF "really didn't cost us much," Mr. Regan was subjected to a veritable barrage of criticism.[5] Senator Don Riegal of Michigan exploded at Mr. Regan, declaring that the 40 percent of the labor force unemployed in the automobile cities of his state could surely use some of that easy money more than the big-city banks it was intended, in his view, to bail out. The hatred of liberal politicians for the lenders in the debt crisis—the big banks—had been slammed on the table; the conservatives' hatred of "communist dictatorships" and wasteful government-run enterprises in LDCs was soon to be expressed as well, creating a vise that squeezed the energy out of attempts to take an enlightened, careful look at the emerging crisis. Flustered by this opening skirmish, the secretary retreated from Senator Riegal's angry outburst. But it was too late. By failing to gauge the sensitivity of the Congress to the issues surrounding the debt crisis, he had blundered into creating a unique coalition of liberals who hated the banks and conservatives who hated the thought of more foreign aid for the big banks' corrupt, communist-inspired borrowers. So swift and overwhelming was the onset of the debt crisis that the government whose own banks were most deeply involved was left playing catch-up ball.

Badly shocked as it was, the U.S. government was not the only one over-taken in 1982 by events of the previous few years. The situation in which the financial world found itself in 1982 was profoundly different from what bor-rowers and lenders had expected as little as a year before. A February 1983 report by Federal Reserve Chairman Paul Volcker to Congress revealed that in 1981 commercial banks had increased their loans to developing countries by $43.2 billion, with $17.4 billion of this coming from U.S. banks: a 20 percent increase over the previous year.[6] Bank lending to LDCs had been growing at this rate since 1975, with lending to major borrowers growing at 25 percent a year. Lending by U.S. banks accelerated even further toward the end of the 1970s, as smaller regional banks increased their participation in international loan syndications. But even during the first half of 1982, when problems with Eastern European loans first surfaced and commodity prices were falling sharply, commercial bank loans to developing countries rose at an annual rate of 12 percent. True, many bankers were growing uneasy about the increasingly large proportion of government financing loans in their portfolios and were thinking of moving toward more corporation financing while being more selective about government loans. Moody's had in March of 1982 downgraded from triple A to double A the bonds of nine leading U.S. banks, leaving only Morgan Guaranty with a top rating. But the bankers whose bonds were downgraded did not seem overly concerned. Harry Taylor, president of Manu-facturers Hanover, was quoted in a *Euromoney* article published in May of 1982 as saying: "It is foolish to suggest that banks are going to stop making general purpose loans, one of wholesale international bankings' fundamental missions in life."[7] Taylor's confidence, not atypical among bankers during the first half of 1982, was soon to be shaken.

By the middle of the year bankers were coming to see a need to reduce the rate at which their sovereign loans were increasing; but this was to be accom-plished at a stately pace, and great care was to be taken not to arouse anxiety among the very best customers that future credits might not be available should they really be required. None felt the need to total up all the credits, particularly to establish how many fell due for payment within a year. Even if this had been done, the presumption of individual banks would have been that rollovers of the short-term credits would be readily available.

But when the figures totaling the obligations of the developing countries, particularly those of their total short-term indebtedness, finally appeared, they caught everyone—bankers, governments, borrowers, and even the IMF—by surprise. The crisis building up from autumn 1981 had until August 1982 struck the bankers only as a warning to contract the growth of lending to developing countries. It took the detonation of the Mexican emergency really to catch their attention; and then they, along with finance ministers and heads of central banks, began to realize that the situation emerging during the fall of 1982 was unique in scope and implication in the history of bank lending to governments.

Crisis Recognized as Unique

Registering the gravity of the situation, the Federal Reserve undertook a full investigation of the exposure of U.S. banks to LDC debts. The results were not very reassuring. Data covering the period through June of 1982 showed that claims on developing countries of the nine largest U.S. commercial banks totaled $30.5 billion—112.5 percent of their assets. On February 2, 1983, this figure was presented, along with broader figures already mentioned, in testimony by Federal Reserve Board Chairman Volcker before the House Committee on Banking, Finance, and Urban Affairs.[8] If all of these loans went bad, the nine largest banks in the United States would be bankrupt. Of course, all of the $30.5 billion was by no means in jeopardy. But if only $10 billion worth were questionable (and Mexico alone owed $13.4 billion), then 36 percent of shareholders' equity in the nine largest commercial banks in the United States would be in jeopardy. For some individual banks exposure was considerably higher, and given fractional reserves and interbank reserve networking, the banking system's vulnerability could be extensive.

In the face of these dangers, banks and developing countries have since World War II acquired a kind of rich uncle in their dealings with one another. Mr. Silva-Herzog's second stop in Washington on that historic August day, the International Monetary Fund, established in 1944, originally had aspirations to be a world central bank that might have included a role as an international lender of last resort. Instead it has become a kind of generous referee that tries to improve communication between banks and borrowing countries when the latter are having some difficulty in repaying loans. The Fund's assets are more in the form of its own communication network between borrowers and lenders rather than in the form of funds sufficient to guarantee hundreds of billions of dollars in LDC debt. In exchange for some adjustment assistance and advice on how better to repay in the future, the IMF requires that its advice, which can include some painful medicine for LDCs to swallow, be followed. But what if stiff IMF adjustment programs—and some will have to be stiff indeed—produce political upheaval and a change in governments? Successor governments might then repudiate the debts of their predecessors as "odious." A careful study of the rhetoric of opponents of IMF-mandated adjustment programs in such countries as Mexico and Brazil suggests that they have read closely the law covering international default.

Still, in the past governments have defaulted on debts to foreign investors on a very large scale without disrupting the international financial system or wrecking a national economy. Why, one might ask, should we expect otherwise from such defaults in the 1980s? The difference lies in the double revolution in lending procedures of the last decade—of magnitude and of locus— with commercial bankers taking LDC debts into their balance sheets instead of distributing them across the whole economy as the investment bankers had done earlier. Procedures followed by investment bankers during the nineteenth and early twentieth centuries and even after World War II meant that government debts were owed to a widely dispersed group of bondholders. When the Bolsheviks defaulted on the czar's debts, the losses were spread all over the world among owners of czarist bonds. Had just a few large banks owned the bonds, they would likely have failed. In sharp contrast, by 1983 the world's commercial banks owned more than half of the over $700 billion or so owed by developing countries—and most of that, as we have seen, was held by a handful of the largest banks. If the big five Latin American borrowers were to default on barely half of their debt, or if investors were willing to value those debts at only 50 cents on the dollar, the impact would be sufficient to bankrupt the six largest banks in the United States, institutions that in 1982 handled deposits of $340 billion, two-fifths of the total demand, time, and saving deposits in U.S. banks. And this does not take into account the many other equally disturbing LDC exposures of these and other banks. Without some major rescue effort, neither the American financial system nor that of the world could withstand a loss of some two dollars out of every five dollars' worth of bank deposits without the massive disruptions in economic activity that have accompanied great financial failures in the past.

Some perspective in early crashes is useful to comprehend the magnitude of the debt crisis threat of the 1980s. Over half a century earlier, between August of 1929, just before the great stock market crash, and March of 1933, when all U.S. banks were put "on holiday" for more than a week, the money stock of the United States fell by more than a third. During that time real income and prices fell by more than one-half and unemployment reached 33 percent. By 1933 per-capita real income in the United States was about what it had been a quarter century earlier, during the depression year of 1908. That depression followed the banking panic in October of 1907,* which had produced the pressure, as much European as domestic, for the development of an American central bank that led to creation of the Federal Reserve System in 1914.[9]

In the United States, the 1929–33 depression was the most severe in its history. Elsewhere the contraction was neither as sharp nor as prolonged, partly because the United States played a smaller role in the world economy then than it does now and partly because the world economy had been more severely depressed than that of the United States during the 1920s and was much less integrated and interdependent than it is now. This was especially true of financial markets. The international activities of U.S. banks were limited. London, not New York, was still the world's financial center.

Some History of the Postwar International Financial System

World War II served to consolidate the position of the United States as successor to Britain in the role of the world's leading economy whose currency was the closest thing available to gold—or to international money. Before and during the war much of the world's gold stock had found its way to the comparative safety of the United States. But that mercantilist indicator of economic strength was validated in the case of the United States by the existence of a great industrial machine, exercised and streamlined by demands of wartime production, and, unlike everyone else's, undamaged in the war. That machine was promptly put to work producing goods in short supply after half a decade of producing mostly armaments. At the same time, the United

---

*Interestingly enough, the panic of 1907 resulted from speculation by some bankers in volatile shares of copper-mining companies. Sharp declines in shares of those companies—like sharp declines in LDC commodity export prices during 1981–82—produced serious problems for a few New York banks.

Dangerous Loans to Princes

States—used to producing and selling most of its output at home and putting domestic matters first again—was looked to, as after the First Great War, to provide leadership in economic matters commensurate with its military and political leadership exercised in wartime. Results were mixed.

After the war the great powers also set to work repairing the world's financial system. At Bretton Woods, a little resort in New Hampshire, where the IMF and World Bank were created during July of 1944 along lines laid down by John Maynard Keynes and Harry Dexter White of the American Treasury, there was a determination to avoid the chaos of the 1930s and create a stable international financial system based on gold and the dollar. In essence, the dollar was tied to gold while other currencies were pegged to the dollar. Countries, in effect, agreed to a plan that put pressure on both surplus and deficit countries to adjust their balance of international payments, although, expecting to be a surplus country for some time, the United States in 1946 effectively nullified provisions that put pressure on surplus countries to adjust. The result was a system biased toward thrusting the burden of adjustment on deficit countries. The corollary was the essentially mercantilist view that since they did not need to adjust, surplus countries were somehow doing the right thing. Asymmetry in the adjustment process was partially the result of an unfortunate lack of American sophistication in these matters; eventually it brought the Bretton Woods system down, but not before it had served for twenty-five years as a loosely organized basis for rebuilding and greatly expanding the world's financial system.

Notwithstanding an inward-looking attitude on formulation of a world financial system, the United States displayed great generosity in helping to rebuild the world's economy after World War II. Between 1948 and 1952, under the Marshall Plan, some $11.6 billion in grants and $1.8 billion in loans were provided to Europe. Another $1.2 billion in grants and loans—most of them grants—went to Japan. Some idea of the magnitude of the Marshall Plan can be gained by comparing the total of $14.6 billion with the $14.0 billion that represented total U.S. expenditure on defense in 1950. By fiscal 1984 U.S. defense expenditures reached about $250 billion, a sum more than double the exposure of U.S. banks to LDC debt and nearly thirty times as large as the $8.4 billion in IMF funding that barely squeeked through the U.S. Congress in November of 1983.

At the end of World War II the United States was like Henry James's Christopher Newman, the central character in his 1877 novel, *The American*:

> His attitude was simply the flower of his general good nature, and a part of his instinctive and genuinely democratic assumption of everyone's right to lead an easy

life. If a shaggy pauper had a right to bed and board and wages and a vote, women, of course, who were weaker than paupers and whose physical tissue was itself an appeal, should be maintained sentimentally at the public expense. Newman was willing to be taxed for this purpose largely in proportion to his means.[10]

American postwar forays into the arena of international finance combined a generosity and naiveté protected, for a time, by immense wealth. But as we shall see, even the wealth, and the mantle of world economic leadership it bestowed, was not sufficient to sustain indefinitely an acknowledgment from other great economic powers that grew up—with much help from Marshall aid—during the 1950s and afterward. By 1960 the dollar's link to gold was strained—to be shored up by a decade of misdirected policies, which—ironically, as we shall see—began to involve American banks in direct lending to LDCs. As always there was tension between the world economy and precious little attention to the economic—as opposed to political—questions it raised for economic policies of the United States. By 1971 Richard Nixon had cut the dollar link to gold, and by October of 1979 Paul Volcker—an American widely and uniquely respected in the international financial community—was virtually booed out of a major gathering of world economic leaders by central bankers and finance ministers disillusioned by a disastrous collapse of U.S. financial integrity.

Volcker's reaction was intense and in proportion to the seriousness of the inflationary prospects that continuation of post-gold, post–oil-shock U.S. monetary policies had created. But the post-1981 snap back toward sound money was faster and stronger than American bankers, LDCs, and leaders of other world economies even believed possible. The result was a 1982 brush with world depression that all—save the weakest, the LDCs—escaped at least for a time. But this gets us well ahead of ourselves—indeed these events, together with other antecedents to the debt crisis of the 1980s, will fill the chapters to follow. For now, let us return to the 1950s.

Eurocurrency Markets Integrate the World's Financial System

The international financial system got a huge boost in the late 1950s from the invention of the eurodollar. The eurodollar is simply a dollar unit of account at a bank located outside the United States, whether British, French, Swiss, German. American banks operating in London or Paris accept eurodollar deposits. Deposits start at $200,000, so it is a market for big players. It got

Dangerous Loans to Princes

started in 1957 when the Soviet government needed a dollar account for numerous reasons of convenience related to the dollar's primacy as an international medium of exchange but, reluctant to deal with an American firm, arranged for one to be started by the London office of the Moscow Narodny bank.

The idea of dollar accounts in non-U.S. banks really caught on during the 1960s when U.S. government restrictions on capital outflows kept U.S. banks from expanding their foreign business save by lending directly to foreigners for periods of less than thirty-six months—a loophole that started U.S. banks making direct loans to LDCs of the very sort that left them so heavily and directly exposed to the LDC debt crisis that emerged two decades later. British and continental banks happily stepped in to fill the gap created by U.S. restrictions on traditional channels for capital flow among nations, and the eurodollar market boomed. The idea came to embrace a "eurocurrency market" of euro-sterling, euro-deutschemarks, and euro-yen although the eurodollar has remained dominant.

Briefly, eurodollars are generated when, say, a Saudi prince receives $25 million for a shipment of oil to Japan. He can direct that the $25 million be deposited at Lloyds bank in London. He may want part of the money to go in overnight while the rest goes in for seven, thirty, or sixty days, depending on alternative uses he may have in mind for the money. Lloyds receives another $25 million to lend, which it likely though not necessarily will lend out in dollars, keeping a tiny fraction, less than half a million, as cash reserves since no reserves are required against banks' foreign currency liabilities to nonresidents. The bulk of the $25 million might be lent to French companies paying for oil or shipments of computers from the United States. The French companies will direct Lloyds to credit their accounts at Credit Agricole. Lloyds has $25 million on its books, Credit Agricole another $24.5 million. The process continues, resulting in creation of a massive volume—about $2 trillion in 1982—of eurocurrency deposits.

The eurocurrency system greatly increased the integration of the world's major financial markets. Major banks in every financial center could offer facilities in any denomination, largely through their eurocurrency operations centered in London. Offshore banking facilities grew up elsewhere, in the Caymans, for instance, and the Bahamas. Any country that threatened to restrict the market saw it simply move away to a country that welcomed the business it generated. When the United States dropped its fruitless restriction of capital flows early in 1974, U.S. banks began to go after more of the offshore-dollar business centered in London and—with the precedent of direct lending established when the controls had first been imposed in 1963—began a rapid expansion of direct term lending to LDCs.

By 1982 the $2 trillion of assets in the eurocurrency system represented a network—highly interdependent and growing more so—of interbank deposits and borrowing facilities. The net size of the eurocurrency system in 1982 was a little less than half of its gross size—around $900 billion. The difference, $1040 billion, was largely accounted for by banks borrowing from each other —for instance, if a bank has $2 million in eurocurrency deposits with $1 million the deposits of another bank, the net addition to deposits within the system is only $1 million. While this interbank borrowing provides much day-to-day convenience, it also means that trouble for one bank entails trouble for all the banks lending to or borrowing from it.

An Interdependent System Is Vulnerable

As with all technological advances that increase the gearing of a system, the development of a eurocurrency system resulted in a highly specialized and interdependent international financial order. It meant that the unique vulnerability of some big international banks to LDC loans created unprecedented vulnerability to those loans for the whole international financial network. Since the system is based on fractional reserves in dollars—banks hold dollars ultimately created by the Fed to serve as a cushion against heavy withdrawals— it places an extra burden on the Federal Reserve System as a national central bank with responsibilities as an international lender of last resort. As the total deposits in the eurodollar system are far greater than dollar deposits in U.S. banks, in the event of a crisis the increased burden might be more than the Fed could manage.

The rapid development of international finance after World War II is something like the explosion of world population during the twentieth century. Although the international banking system has been around for a long time —since 1960 at least—its size has grown so fast that it dwarfs anything that had gone on before. At the same time, the exposure of commercial banks to economic and political risks has grown to a point where previous exposures seem almost inconsequential—a phenomenon distinct from the specter that haunted prophets of doom in the 1960s, who saw the world financial system collapsing under the wild gyrations of unpegged exchange rates or the unimaginability of a world financial system without a fixed price of gold. Those earlier fears grew out of a confusion of cause and effect in the interpretation of the 1930s experience. Beggar-thy-neighbor exchange rate policies and departures from gold were results of the Great Depression, not its cause. Between August 1971 and March 1973 the international economy moved away from fixity of exchange rates keyed to a preestablished dollar price of gold; moved away out of necessity, not as a result of any conscious design or decision by governments. Thus was the international system transformed into one that, as it turned out, was far better adapted to accommodate oil shocks and the massive capital flows among the ever-more-integrated financial markets of the world.

The fears of the world economy being ruined by exchange rate fluctuations were speculative and vanished in the light of common day. Such disorders might conceivably have increased the risk inherent in international trade in goods and securities, but the computer, the telex, and the growing forward markets absorbed or obviated most shocks. But in the case of the LDC lending explosion, very real assets have been transferred and must be paid for: if not by the ultimate borrowers, by whom?

In the cold light of the post-1982 period, when the music that played for almost a decade to a tune that saw accumulation of $700 billion in LDC debts has stopped, it is easy to claim that the institutions that were prepared to list hopes and assumptions as collateral logically should make up the difference. But their very capacity to make these loans reflected the vast range of their services; to liquidate such subtle and far-ranging instruments might deal an intolerable blow to the world economy. The threat to these key banks' existence is a far greater danger to the world economy than was the possibility that certain ministries' or investors' or speculators' fingers might be pinched by day-to-day changes in the price of one money in terms of another or in the dollar price of gold.

Comparing the aftermath of the oil shocks and the debt crisis, it is clear that economic systems are well adapted to dealing with price changes. Indeed it is price changes that signal the need for adjustments—saving energy—in the way economic activity is conducted. But at the same time we have learned that economic systems are less well adapted to making accurate forward judgments, especially, as in the case of the debt crisis, where the value of experience has been written down to zero. While it is theoretically possible to insulate them from the shock of major losses on bank assets, fractional-reserve financial systems have in practice been vulnerable to such shocks. Bank losses damage the confidence of depositors, who try to protect themselves individually by trying to withdraw their funds from the banks and thus without meaning to, raise the likelihood that bank depositors as a group (including themselves) will lose their money with the insolvent bank. A confidence crisis creates a liquidity crisis, which, if not addressed with injections of new liquidity, results in a solvency crisis. There is always the difficulty of perceiving soon enough the fallacy of composition whereby if one depositor tries to withdraw his money from a bank, he succeeds, while if all depositors try, all will fail. Recognizing this, economists—notably Milton Friedman and Anna Schwartz—argue that the Great Depression need not have occurred had the Federal Reserve been willing to serve as lender of last resort. Later in 1933 the creation of the Federal Deposit Insurance Corporation lessened the probability of a repetition of such a chain reaction of disaster in the United States by assuring that smaller depositors need not fear losses from a bank failure and thereby obviating panics.

Financial Vulnerability in the 1930s

It is chastening to reflect that after fifty years' experience, the function of an international lender of last resort is less well developed in the 1980s for the international financial system than it was for the United States during the early 1930s. During that disastrous epoch U.S. banks discovered the hard way that neither legal reserves required by the Federal Reserve nor the presumed availability of a lender of last resort was of much help when trouble began. The process of deterioration accelerated after the failure of the Bank of the United States on December 11, 1930. It had over $200 million in deposits, little now but quite substantial in the poor and uninflated days of 1930, and as Friedman and Schwartz point out: "Though an ordinary commercial bank, its name had led many at home and abroad to regard it somehow as an official bank, hence its failure constituted more of a blow to confidence than would have been administered by the fall of a bank with a less distinctive name."[11] Over the two years following the failure of a bank whose name and aura are echoed today by the Bank of America, which with its $94.3 billion in deposits is the largest U.S. bank, there evolved a liquidity crisis whereby bank deposits reverted steadily back into currency. Banks were forced to hold more and more reserves, producing a one-third drop in money stock between 1929 and 1933 and ultimately a general panic when deposits were flowing out faster than the banks could obtain the currency demanded by depositors. The culmination was the proclamation by President Roosevelt—aimed at avoiding a full-scale solvency crisis—of a nationwide "banking holiday" at midnight on March 6, which closed all banks for over a week.

The panic had stemmed from rumors of large loan losses accentuated by fears—that proved to be justified—that the new Roosevelt administration might devalue the dollar. This occasioned speculative accumulation of foreign currencies and gold. On January 31, 1934, the President devalued the dollar by almost 40 percent against gold—from $21.65 per ounce to $35. It remained at that price, which, by a curious reversal, changed its nature from being a concession of economic weakness to a sign of American economic preponderance, for almost forty years. And during those forty years after 1934 the dollar itself lost three-quarters of its value, a sign to those who could see that for a generation the world was more on a dollar-exchange than a gold-exchange standard.

Friedman and Schwartz attribute the panic that preceded devaluation of the dollar in 1933 and its disastrous consequences to the failure of the lender of last resort: "The central banking system, set up primarily to render impossible the restriction of payments by commercial banks, itself joined the commercial banks in a more widespread, complete and economically disturbing restriction of payments than had ever been experienced in the history of the country."[12]

Parallels: 1930s and Later

Comparison of the situation facing the U.S. and world financial systems today with the situation during the Great Depression reveals two basic truths. Today we understand better how technically to deal with the problems, like banking panics, that can arise from bad loans. But the scope of the problem and the inability of banks to have recourse against their major international debtors are unprecedented. The latter truth is a manifestation of the increasing involvement of the United States and its banks and businesses in the world economy together with a reduction in the country's relative economic size. These developments became particularly noticeable at the start of the 1970s. Since then both new international opportunities and new international challenges have been cropping up steadily, especially as world financial markets have become more closely interconnected.

The extraordinary exposure of institutions around which the world financial system is organized—and hence of the system itself—to developing country debt came about as part of the fundamental change in the world's monetary functions initiated by the United States on August 15, 1971, when President Nixon cut the international link, set thirty-seven years before, at $35 per ounce between the dollar and gold. Until 1971 the U.S. government had sought actively to discourage bank lending abroad for fear that the outflow of dollars could jeopardize the $35 ratio, which was ever more sacrosanct as it lost empirical justification and raised the specter of currency prices being set by the market. A fixed-dollar value of gold was at the heart of the Bretton Woods international monetary system. With the gold-dollar link broken, the U.S. could relax and enjoy what its currency bought in the world marketplace, and its money center banks began an ever more vigorous expansion of their international lending operations.

By 1974 all the forces set in motion during the previous decade met, and in Churchill's phrase about the Treasury mind and that of the last fiscally conservative Socialist, "embraced each other with the fervor of two long separated kindred lizards." The big banks, freed of any restrictions on direct lending abroad and facing in 1974–75 a sharp recession at home, were attracted by the prospect of a new economic order associated with the rapid development of the resource-rich economies of Latin America and Asia. When, during 1973–74, the Organization of Petroleum Exporting Countries (OPEC) quadrupled oil prices the machinery was in place to create, almost overnight, an enormous increase in the potential for international lending. The oil exporters had enormous new funds, which they deposited in the world's major commercial banks. The oil importers, many of whom were still viewed as promising aspirants to the new economic order, sought loans from the deposit-choked banks to pay for their oil imports. If the world's largest commercial banks were to survive as industry leaders, they had to show themselves able to seize their share of the massive "recycling" required by the oil boom: and indeed, the newly dynamized big bank sector was out to take more than its traditional share. The need to borrow was more than matched by the pressure to lend the $68 billion combined surplus of the oil exporters during 1974. Frightened treasuries stood beside overexcited journalists to crow up the new dawn. And lending continued to rise at an unabated rate even when the combined surplus of the oil exporters fell to about $35 billion in 1975. Surely this was, in a contemporary phrase, "a blip on the radar" that went unheeded.

It is in the nature of things that deposit inflows and outflows are far more volatile than are investment opportunities. Compounding this problem for banks is the need to employ the funds, particularly in large and profitable chunks, in order to earn returns competitive with other banks' rate of return. Nor is this problem unique to the twentieth century. Writing of the Medicis' heavy losses during the fifteenth century to King Edward IV, the historian of their bank opines negatively: "Rather than refuse deposits, the Medicis succumbed to the temptation of seeking an outlet for surplus cash in making dangerous loans to princes."[13]

Dangerous Loans to Princes

Recycling

There is little to be gained from looking backward with disapproval at the consistency of human folly except to notice how each generation thinks itself immune to its predecessor's mistakes. It is more useful to understand how the world appeared to those looking ahead when petrodollar deposits began to flood the big-time banks so rapidly in 1974. First, some lending to overcome cyclical imbalance is always justified. As Keynes's generation sagely observed, perhaps too often, one may perish of thrift as of profligacy. Second, governments were exhorting banks to undertake their recycling responsibilities and heaping praise on those that moved most rapidly. The emphasis was on "need" and a "new economic order," with an eye on political forces playing upon many of the borrowing countries. Particularly compelling were the amounts of money involved. Banks lent about $25 billion to non-oil developing countries during 1974 alone. Another $10 billion was lent to these countries by governments and international agencies, money that served as implicit approval for the acceleration of bank lending to oil importers. Twenty-five billion dollars in loans meant $150 million in bank earnings for every 1 percent in the interest differential between the bank as borrower and the bank as lender. Combined, these fees could reach half a billion dollars annually if spreads and service charges totaled slightly over 3 percent. By the end of the decade over half of the profits of major commercial banks—assets of the top fifteen totaled about $1.2 trillion in 1982—came from international business.

And remember also that this first surge of petrodollar lending seemed remarkably successful. The banks forged an initially most profitable working arrangement with those governments of developing countries prepared to serve as intermediaries between foreign banks and ultimate Third World borrowers. So the banks came to believe that they were being spared the tremendous expense of ascertaining the viability of each separate industry, not to mention huge start-up costs—who speaks Swahili at Chase Manhattan?—and were thereby able to place—oh, quite safely—huge sums of OPEC funds with a single loan to the government representatives of local business interests. The borrowers in developing countries believed that they had discovered the advantages of borrowing very large amounts at once rather than seeking funds related to any particular project, whatever its economic merits may be. Or, perhaps, since the rulers of such countries are more noted for traditional shrewdness than modern economic acumen, they merely remembered the words of the great Lord Curzon: "Presumption is the secret of success!"

And these beliefs crystallized just when a surge of liquidity made the banks as lenders particularly anxious to place large amounts of money, by bankers' standards, very fast.

Inflation Validates the First Recycling Wave

At first, performance on the loans made during the mid-1970s under the newly forged and mutually beneficial partnership between banks and developing countries was quite satisfactory. This was not an index of their economic soundness but of the industrial countries' decision to inflate their way out of the first oil crisis, which worked to validate the sustained increase in the price of petroleum. Inflation in industrial countries rose from a 1963–72 average of 4.2 percent to 11.7 percent in 1974 and 11.1 percent in 1975, dropping back to an average of 7.7 percent between 1976 and 1979.[14] An important consequence of this greater-than-expected surge in inflation was a very low, and indeed sometimes negative, real cost of borrowing after inflation had been subtracted from interest rates. Bankers in the mid-1970s who expected a resumption of 4 or 5 percent inflation and were therefore willing to lend at 7 or 8 percent got burned by escalation of prices at or above that level. As a result they learned to tie rates on LDC loans to market rates or, failing that, either to lend short or to build a healthy premium for inflation into their lending rates.

Besides benefiting from the effect of inflation of their borrowing costs, developing countries saw rises in the prices of many of their exports relative to prices of their imports; the "terms of trade" improved markedly, even for oil importers during 1976 and 1977. Of course, inflation accelerated in developing countries at the same time. For oil exporters awash in new revenue, it rose from a 1968–72 average of 8 percent to 18.8 percent in 1975 and 16.8 percent in 1976. For oil importers, inflation rose from a 1968–72 average of 9.1 percent to over 27 percent in 1975 and 1976, fueled both by worldwide inflation and by the sharp increase in liquidity caused by heavy capital inflows: the loans that flowed in did not all flow out to pay for oil, especially in the developing countries.[15]

Accompanying the steepening inflation were growth rates somewhat below 1968–72 averages for all developing countries. By 1978 average growth for oil exporters fell to 1.8 percent, while it held fairly well at 5.5 percent for non-oil developing countries before dropping to 2.8 percent in 1979. During the middle and late 1970s developing countries combined sharply higher inflation rates and lower growth compared with previous years' performance—the ominous repetition of a pattern that had emerged for developed industrial economies as well.

And yet a more ominous part of the response to the first oil crisis was its setting the stage for the second oil crisis, which, in turn, predetermined a new, more complex, and deeper cycle of disaster. By 1979 the general inflation that followed the 1973–74 jolt had sharply reduced the "real" (inflation-adjusted) price of oil. Prices in industrial countries where oil exporters bought many of their imports nearly doubled between 1973 and 1979; and in developing economies inflation had more than quadrupled prices. By 1978 oil exporters' terms of trade had begun to deteriorate sharply, falling during that year at a rate of 10.7 percent, while terms of trade for industrial countries improved simultaneously at a rate of 2.7 percent. OPEC members did not have to be economists to see that the developed West was paying with the right hand and taking back with the left. They responded.

Second Oil Crisis: More Borrowing but Less Inflation

During the period from mid-1979 to the end of 1980, OPEC nearly doubled oil prices again. By the end of 1981, aided by the sharp curtailment of Iranian oil and a final spurt of money growth between mid-1980 and the first quarter of 1981 in the United States, oil prices had gone well beyond that. From levels of $2.9 billion and $39.2 billion respectively during 1978, the combined current account surplus of oil exporters and the combined current account deficit of non-oil developing countries rose in 1980 to $116.4 billion and $86.2 billion. By 1981, as in 1975, the ratio had been reversed, with a combined $68.6 billion surplus for oil-exporting developing countries compared with a $99 billion combined deficit for oil-importing developing countries.[16]

Informed opinion braced itself for a return of the explosive inflation that had followed upon the price surge of 1974–75; many contracts were written in money, commodity, and labor markets that merely quantified what, at the time, seemed intelligent anticipation of just such an updraft. The year 1981 saw the largest increase ever in debt to banks of non-oil developing countries: $43.3 billion piled on top of a $40 billion increase in 1980. Interest rates were high on this newly contracted debt; but surely, everyone thought, looking back on experience of the last two decades, most real interest cost would be largely eliminated by inflation or made bearable by sharply higher prices of goods sold by borrowers paying the high rates. The gross interest payments of oilless developing countries nearly doubled, from $28 billion in 1979 to $55 billion in 1981. Over the same period debt service as a percentage of exports rose from 21.9 percent to 22.9 percent, and more sharply, to 28 percent in 1982.[17] Like the White Queen, the non-oil developing states were having to run as fast as they could to stay in the same place.

One reason that their debt service seemed not to rise sharply relative to exports is that published statistics exclude short-term loans (those made for less than a year). Though data are scarce, it is clear that such short-term loans greatly expanded during 1981 and 1982. The Morgan Guaranty Trust Company estimates that such borrowing by developing countries passed the $130 billion mark by the end of 1982. This expansion joined the sharp run-up in debts contracted on terms anticipating renewed high inflation during the 1980s to collide with a serious program of inflation control initiated by the United States in the spring of 1981. The result was a major economic crisis for the world during the following year.

LDC Debt Crisis Signals a Global Debt Crisis

Characteristically, the United States, with its eye squarely on its own formidable inflation problems, failed adequately to take account of the international —not to mention domestic—repercussions of actually undertaking, in the spring of 1981, a serious program of inflation control after more than three years of falsely claiming that such a program was in place. The weakest link in the almost decade-long inflationary bonanza, rapid growth in lending to LDCs—much of which was going to support continuation of a consumption binge—snapped. But the familiar cycle, whereby some economic growth leads to hopes for a better future that are prematurely legislated into promises by well-intentioned legislators who subsequently must borrow by printing more money to levy the inflation tax when growth falls short of expectations, was not confined to developing countries. In the United States Ronald Reagan's willingness to cut taxes was not adequately matched by a willingness to cut back on promises, especially those to us and our allies for the security implied by a preponderant U.S. military machine. Nor, true to his promise and much to the surprise and real regret of many who had to act pleased, was he willing to continue levying the inflation tax. The result was a surge of red ink that spilled across the President's supply-side manifesto and its wishful promise that somehow cutting taxes would release enough pent-up energy to increase the total tax take. The $200 billion or so a year that this wishful thinking, the stuff of which all crises are made, was adding to the American debt—more and more of which was being bought by foreigners flush with dollars earned by selling goods to Americans whose purchasing power outside their borders was boosted sharply by a dollar made strong by the high interest rates required to place $200 billion atop the world's growing mountain of debt—meant that the LDC debt crisis was no more than a speeded-up version of the world's debt crisis. And that debt crisis, combined with the hopes that had been legislated into promises in Europe, Japan, and elsewhere, meant that the developing world's debt crisis that surfaced after the United States had for a year been draining the world's reservoir of liquidity was only the tip of the iceberg. Beneath the surface lurked a huge, increasingly indigestible chunk of debt, and neither the higher taxes nor the lower spending needed to melt it down seemed imminent.

In short, the term debt crisis carried a double meaning during the eighties. The LDC debt crisis represents the acute phase—the leading edge, if you like —of a much larger crisis whose elements are always the same: consistent spending in excess of revenues, which results first in accumulation of debt followed by inflation or outright default. The latter is the only option open to LDCs, much of whose debt is indexed and largely protected against inflation. This is not, however, the case for industrial countries.

This process, as we shall see in chapter 2, has been repeated countless times during the eight centuries that governments have borrowed to finance their activities. What varies are the stories told to sustain the process—the "something different this time"—be it oil, "the country of the future" image of Brazil, or the magic of supply-side economics where somehow tax rates become a major factor of production. If they are believed long enough, and they usually are, the eventual result is one of the financial crises that punctuate history with remarkable regularity at intervals of about fifty years; two generations seem sufficient effectively to erase from mankind's memory the lessons of the past.

Crisis Worsened by Earlier False Starts on Inflation Control

As already has been suggested, after the second oil crisis the U.S. Federal Reserve System elected not to meet the outflow of money to Arabia simply by pumping in new, inflationary dollars, fearing that to repeat the policies of the 1970s would launch a devastating price spiral which might blow the economy away. However, very few observers—including apparently the big banks and the developing countries—expected that the Federal Reserve would really follow through. They can perhaps be forgiven: the Fed had been promising to control inflation since the late 1960s when the burden of financing the guns and butter of the Vietnam war and the Great Society had produced the first major, sustained depreciation of the value of money in the United States since World War II. Based on the Fed's longstanding tendency to cry wolf when it came to containing inflation, recovery from the 1981 recession was confidently expected by the end of that year. It never arrived in 1981. Nor did it arrive during 1982, when hopes for recovery dissolved like a mirage with the arrival of each month or quarter that had been designated as the date for resumption of healthy growth. Countries that had borrowed heavily to pay for more expensive oil saw their exports drop in both price and volume, began to borrow short term to finance "temporary" shortfalls, and when these persisted continued to borrow yet more short term in order to service long-term debts. In less august banking circles, this is taken as a sign that one has lost one's principal. When the time came during mid-1982 to renew such dangerous loans, the banks finally balked. And so it proved that governments were after all mortal and could miscalculate and run out of cash even as car buyers and shopkeepers do. First Mexico began to miss interest payments on its loans, then Venezuela, Nigeria, and even the "miracle" country, Brazil. The world debt crisis of 1982 had begun.

The System Is Fragile

The finance ministers and heads of central banks confronted by LDC debts and the state of the world economy late in the summer of 1982 were typically in their fifties or sixties. Paul Volcker was fifty-five; Donald Regan, sixty-four. Jacques de Larosière, managing director of the IMF, was fifty-three. As children and adolescents these men had lived through the depression. As young men they had endured the horrors of World War II, legacy of the economic chaos that followed from the punitive peace settlement after World War I. They knew that the fabric of international systems can and does tear and that repairing it can be a very slow, costly, and difficult process. They had seen the price exacted from the neighbors of their childhood, the fellow-soldiers of their young manhood, greater prices than those asked of policy makers.

Looking at the situation they faced, as they prepared over the weekend at

Dangerous Loans to Princes

the key Interim Committee meetings for the start on Monday, September 6, of the 1982 Annual Meetings of the International Monetary System and World Bank, they could be forgiven their deep foreboding.

The world financial system had never really been severely tested on its ability to prevent such banking problems as emerged in the United States under Herbert Hoover from developing into a crisis like the one that forced the 1933 bank holiday. The existing international framework for dealing with crises was really no better developed than the Federal Reserve System had been in the early 1930s. There was and is no comprehensive system of deposit insurance outside the United States to prevent bank runs. While the Federal Reserve System takes seriously its function as lender of last resort for the United States, there are many areas—for instance the $2 trillion eurocurrency markets—where such responsibilities are not clearly established. Nor is any institution likely to come forward to seek such responsibility just now. Notwithstanding their impressive names, the International Monetary Fund and the World Bank are short- and long-term international *lending* institutions only. They possess little, if any, of the powers or resources that would be required of a true world central bank.

Writing about the infancy of the international financial system just after World War I, Keynes described its future prophetically:

Before the middle of the nineteenth century no nation owed payments to a foreign nation on any considerable scale, except such tributes as were exacted under the compulsion of actual occupation in force and, at one time, by absentee princes under the sanctions of feudalism. It is true that the need for European capitalism to find an outlet in the New World has led during the past fifty years, though even now on a relatively modest scale, to such countries as Argentina owing an annual sum to such countries as England. But the system is fragile; and it has only survived because its burden on the paying countries has not so far been oppressive, because this burden is represented by real assets and is bound up with the property system generally, and because the sums already lent are not unduly large in relation to those which it is still hoped to borrow. Bankers are used to this system, and believe it to be a necessary part of the permanent order of society. They are disposed to believe, therefore, by analogy with it, that a comparable system between Governments, on a far vaster and definitely oppressive scale, represented by no real assets, and less closely associated with the property system, is natural and reasonable and in conformity with human nature.[18]

Immediately following this paragraph, Keynes wrote: "I doubt this view of the world."[19]

The system is still fragile. It remains to be seen whether Keynes's doubts, so grimly justified by the 1930s, will again, fifty years, later prove true.

Chapter 2

Treaties of Tyrants

The sovereignty of peoples is not bound by the treaties of tyrants.

Statement by the post-revolutionary government in France repudiating debts of the royalist regime, 1792.

THE MOST remarkable thing about government debts is the consistency with which they are repudiated by war, inflation, simple fiat, or the disappearance or reconstitution of the government that issued them. The tendency of governments to overextend themselves and the ultimate requirement for higher taxes to pay for rapidly growing obligations was clearly recognized by the first moneylenders. When the Italian city-states borrowed money from syndicates of moneylenders within their citizenry rather charmingly called *monti* ("piles" [of money]), these latter typically furnished their loans in exchange for the security of the city's future tax revenues, usually, in fact, taking over the collection of such imposts themselves to remove from the city any temptation to use the revenues for further good works or war on the usual politician's optimistic assumption that the funds to pay the *monti* could be borrowed elsewhere: a game of musical chairs that always ended in default, as the victims of John Law and Carlo Ponzi were to discover centuries later. There was only one stream of future tax revenues, and the *monti* who lent to a volatile, medieval Italian city-state wanted to be sure that it was used to repay their loans as agreed.

Treaties of Tyrants

Early History of Government Borrowing

The first bank loans to governments such as those by the *monti* to Italian city-states during the twelfth and thirteenth centuries were distinguished by two features. Governments were borrowing from their own citizens organized as an influential syndicate, and the borrowing was collateralized by tax revenues, usually collected directly by the lender instead of the borrower. In contrast, Citibank has no explicit claim on the tax revenues of Mexico or Brazil and surely is not involved in their collection. Even if such a claim existed, it is doubtful whether it could be enforced by law since Brazil is not borrowing from its own citizenry and falls outside the jurisdiction of U.S. courts. Besides, as we shall see, there exists no coherent body of law governing the relationships of borrowers and lenders across frontiers.

The critical aspects of a loan, from the lender's point of view, are the source and mechanism of repayment and the nature of remedy in case of default. The *monti* could minimize the second possibility by collecting the earmarked taxes on their own. Since they were physically present and influential in the community being taxed and therefore in a good position to judge its ability to pay, they were typically able to avoid defaults. But as their resources grew they began to look outside their own borders for profitable investments, much as their successors, choked with petrodollars, were to do centuries later. The propensity of early monarchs to pursue wars for territory and sources of additional tax revenues provided them with ready customers.

Late in the thirteenth century Edward I of England turned to the bankers of Florence, the Bardi and the Peruzzi, to finance his wars. The tax on wool exports was offered as collateral, apparently with no thought given to the perversity of offering as collateral a measure that reduces the ability to repay the loan; a tax on exports could not help. Perhaps due to the burdensomeness of this arrangement or perhaps because he felt able to squeeze enough out of his own citizens to pay for his wars, Edward III defaulted on the Italian debts shortly after donning the English Crown. The Bardi and Peruzzi banks collapsed, setting back Italian banking for a generation.

Edward was mistaken in his belief that his own citizens could bear the full costs of his royal enterprise. He was forced to turn to the moneylenders of Brussels who, chastened by the experience of the Italians, required him to turn over his crown as security. By 1340 Edward was forced to petition Parliament for additional funds from the Estates of the Realm, arguing that the only alternative was for him to surrender himself as hostage to secure his royal debts.

The experience of the English kings during the thirteenth and fourteenth centuries with their European creditors was unique in a number of ways, in addition to representing the first instance of international lending by the forerunners of modern banks to the forerunner of modern governments. Initially the notion of collateral so thoroughly established by the *monti* seems to have been ignored by the Bardi and Peruzzi. The Florentine bankers, like their twentieth-century counterparts who moved to Rio and Mexico City, did move to London, no doubt to oversee as closely as possible the collection of the hypothecated tax on wool exports. But they were not in a position to collect the tax themselves, being, after all, foreigners in London without the pervasive influence exercised by the *monti* over the affairs of their own city-states. Although their loans were to the Crown of England, the head upon which that crown rested made a difference. In the first year of his reign Edward III repudiated the Italian debts of his predecessor. The issues of collateral and succession, or the responsibilities for debt in the event of a change in government, arose very early in the history of international bank lending to governments and were found to be very awkward matters to resolve.

The insistence by the moneylenders of Brussels on the physical possession of Edward's crown as collateral was a crude yet effective device that provided them with an object of considerable tangible value as well as with a literal sign that their claim was on the Crown of England and not just on a particular king.

We can readily distinguish the experience of the Italian city-states, and of Florence in particular after the English default brought down the Bardi and Peruzzi, from the likely consequences of a contemporary government default on liabilities to foreign banks. Earlier the losses were almost entirely absorbed by the ruined bankers—the houses were partnerships, not joint-stock enterprises. Viewed in modern terms, it was the owners or shareholders of the banks who absorbed the loss since the capital was their own. The practice of issuing bank notes (contemporarily checking deposits) based on reserves of the bank had not evolved and therefore the issue of losses by depositors—as distinct from shareholders—did not arise. Money issued by banks hardly existed prior to the fifteenth century since monarchs jealously monopolized the right to issue money as a source of revenue. That revenue, called seignorage, arose from the exclusive right of monarchs to mint coins that cost less than their face value to produce. Debasement—assigning a value to the coinage greater than its gold or silver equivalent because the king says so, and keeping the difference because he needs it—was the forerunner of government printing unreal money to pay its real debts because we perceive ourselves unable to do without all that government does while we refuse to pay directly all the necessary taxes.

Establishment of the Bank of England: Shareholders and Depositors

Ironically, it was monarchical profligacy that ultimately resulted in the loss of the monopoly on money creation by the English Crown. The result was the creation of what became a great financial power, the Bank of England, which operated as a private bank until after World War II and which, like all modern banks, dealt both with shareholders (owners) and depositors (customers). In 1694, after years of war with the French, King William of England a not very popular, revolution-installed Dutch prince, figuring whereas an earlier English government had run up plenty of debt, had taxed, mortgaged, and cajoled through a state lottery all that he could to pay his ongoing expenses in a desperate struggle with Louis XIV. Even with these efforts, William, together with his English predecessors, had managed to run up a debt of 20 million pounds sterling, more than ten times the Crown's annual revenue. For a loan of 1.2 million pounds sterling William relinquished to the Bank of England the six-century monopoly of centralized coinage held by the English monarchy since William the Conqueror. Though bank-managed money had existed in the burgher-ruled states of northern Italy and the Low Countries since late in the fifteenth century, the Bank of England had in 1694 purchased the right to issue notes that circulated right along with the coinage of the monarch. It took some four years after the Bank was established to get the old, debased money out of circulation and replace it with a new, solid coinage, but by 1698, after a recoinage had restored England's money to the Elizabethan silver standard, it was the Bank of England, representing the men of influence who owned it, that controlled the value of England's money. And as shareholders who owned the bank, the short-run interests of these men did not always coincide with those who kept their money on deposit.

The creation of the Bank of England firmly established the idea of a bank that was more than just an intermediary between borrowers and lenders. Its depositors, who were not generally among its owners, received notes or claims on the bank that served as a means of payment. The Bank of England could create money, and that meant its failure or any possibility of its failure could affect the economic well-being of many besides its owners, a fact they clearly recognized by tying its note issue to the quantity of available gold. This quantity fell as notes became too plentiful relative to the purchasing power of gold outside England: gold flowed out, lowering the note issue, and thereby restored its purchasing power. Conversely, a shortage of notes attracted gold from abroad and expanded the note issue. This was the "specie flow mechanism" described by David Hume in 1752 (which, considering that Hume had made his reputation demonstrating that we can never adequately establish the relation of cause and effect, has a charm all its own).[1]

Emergence of International Banking

Prescient as they were, these episodes and insights occurring between the twelfth and mid-eighteenth centuries existed apart from a coherent world economic system. We can talk of such a coherent system perhaps from the late eighteenth century. The first truly international crisis involving recession and a little repudiation began in 1837. The worst—and the most predictable—had occurred after the 1929 smash in the United States brought down the war-strained economy of Europe. Beside predating perhaps the deepest and broadest world economic crisis, the decade of the twenties was the first time that a storm had hovered for so long before it broke. In the early eighties the possibility of government default for the first time assumed the same dimensions—but much more of this later, as it is our main story.

Returning to our chronology, by the nineteenth century the Bank of England had nearly evolved into a modern central bank with its notes serving as backing for outlying banks and an ability to serve as a lender of last resort. The strict linkage of its note issue to the quantity of gold currently on hand had—as a forerunner of more serious slippages to follow later—been modified by the use of the bank rate on interest paid to foreign depositors. A rise in the bank rate would cause foreigners to ship gold to the Bank of England in exchange for interest-earning deposits. In his classic account of the London money market *Lombard Street,* Walter Bagehot opined that 7 percent would bring gold to London from the North Pole while 10 percent would bring it from the moon.[2] If those returns seem low by contemporary standards, it is because the payments were largely "real," needing to take little, if any, account of inflation: the Bank of England was not free, as are modern central banks, to depreciate its liabilities by printing unlimited quantities of money. Any such action by the "Old Lady of Threadneedle Street" would have driven gold out of England and eventually required the country either to contract its note issue or pay higher interest on its deposits than it earned on its investments.

The evolution of the Bank of England produced a bank that borrowed and lent internationally and had depositors who were distinct from the owners of the bank. The notes held by those depositors came to be used as part of the circulating money supply of England and as a means of financing international trade and investment. By the nineteenth century the well-being of what was by then the international financial system was tied to the operations of a commercial bank operating on fractional gold reserves. But like those of most other banks of the time, the international operations of the Bank of England were largely centered on very short-term lending to finance the day-to-day turnover of trade; manufacturing investment was still the business of individual speculators. British exporters who had shipped goods abroad would take their "bills," or accounts receivable, to the various discount houses that would exchange Bank of England notes (or their own notes convertible into those of the Bank of England) for the accounts receivable. In that way, the exporter with, say, 1,000 worth of accounts receivable payable in ninety days by importers abroad could convert the account into 995 worth of cash with the "discount" of 5/1000 per quarter. When he received the full 1,000 from the importer, he paid it over to the discount house, yielding that institution an annual return of about 2 percent.

International Lending Before World War I

Nineteenth-century banks did not engage in large-scale direct lending to foreign governments, but governments did frequently borrow from the public at large through the investment bankers. The vehicle for such borrowing was typically the issue of bonds by the borrowing government. Bankers like the Rothschilds and Barings in London and J. P. Morgan in New York placed the bonds with investors. The collateral was once again the ability of such governments to collect taxes, although, once again, lenders had little power to oversee, let alone to expedite, their collection.

The vulnerability of international lenders to default on uncollateralized loans to foreign governments was as apparent in the nineteenth century as it had been five centuries earlier when Edward III shrugged off his obligations. British investors with large quantities of available capital purchased bonds issued by a number of the American state governments. The option of outright repudiation was exercised by, among others, Mississippi, Maryland, Pennsylvania, and Louisiana. The British Council of Foreign Bondholders formed in 1868 continues in the 1980s to seek compensation for losses suffered in the states' repudiations during the 1840s. Perhaps echoing the widespread losses by nineteenth-century Britons on their American investments, Dickens has his miserly Scrooge experience a nightmare in which piles of his solid British assets become "a mere United States' security."[3] The echo recurs in a 1983 newspaper cartoon where disappointed bank robbers who have managed to break into a bank vault complain: "Some bank job! Nothing but IOU's from Brazil and Mexico."[4]

Despite the losses of British investors in North America during the nineteenth century, international lending expanded steadily. During the half century before 1914 the total value of foreign investments by big creditors rose from just under $4 billion in 1864 to about $44 billion in 1913—about $44 billion and $480 billion respectively in 1984 dollars. Expansion of international lending at an average rate of almost 5 percent a year over a fifty-year period resulted in huge accumulations of foreign debt in big creditor nations. The biggest lender during this period was the United Kingdom. The 4 percent of its national income—American banks invested just over 3 percent of gross national product (GNP) abroad in 1981—and 40 percent of its gross capital formation that went overseas during the prewar period resulted in British ownership of 47 percent of total world long-term investment, with France and Germany contributing 20.7 percent and 16.1 percent respectively to the remainder of holdings. On the other side of the ledger, the United States was a net debtor up until the outbreak of World War I, more so perhaps than aggregate figures show since New York money invested in Wyoming was almost as much a foreign investment as French money in a Ukranian railway.[5]

The funds generated principally in Europe were distributed all over the world. About 27 percent stayed elsewhere in Europe—with a big share of that portion going to Eastern European countries and Russia—about 20 percent went to Latin America, 15 percent to the United States, 14 to Asia, and the rest to Africa and Canada. Most of the lending was accomplished through purchases of fixed-interest government bonds or railroad shares (and railroads in such countries usually had so much government support, financial and political, that the distinction was largely nominal). The growing international flow of capital was accompanied by expansion of world trade at unprecedented rates, especially during the latter half of the nineteenth century, which saw decadal growth rates of over 50 percent. Early in the twentieth century trade continued to expand at a rate of nearly 40 percent per decade.

Defaults After World War I

The effects of World War I on international lending were as profound as its effects on European society. For better or worse, a way of life disappeared. Investors in Europe faced devastating losses both through forced sales of assets to pay for the war and through default or repudiation in the politically unstable climate created by war's turmoil. Germany lost virtually all of its foreign holdings. Those not sold during the war were confiscated at the peace settlement. Britain lost 15 percent of its foreign holdings, which meant loss of about 6 percent of its gross capital holdings—equivalent to about two years' worth of England's total output.

France, among the "victorious" allies, was the biggest loser on foreign investments, which were concentrated in Russia. In 1918, less than one year after coming to power, the Bolshevik government in Russia repudiated all debts that had been contracted by the czarist and Kerensky governments. The decree issued by the new Soviet government was simple and unequivocal: "All foreign loans are hereby annulled without reserve or exception of any kind whatsoever."[6] In issuing its unilateral decree of repudiation, the Soviets simply rejected the claim of the creditor countries that it had an obligation to foreign bondholders to honor debts of previous Russian governments. As a basis for rejection the Soviets stated that "governments and systems that spring from revolution are not bound to respect the obligations of fallen governments."[7] French investors, with about one-half of their foreign holdings in Russian bonds, lost an estimated $4.5 billion—about 45 billion 1982 dollars—as a result. In addition to devastating financial losses—which included the lifetime savings of many middle- and upper-class French families—the French had to endure the citation by the Bolsheviks, as part of their rationale for repudiation, of the 1792 statement by the post-revolutionary government in France that "the sovereignty of peoples is not bound by the treaties of tyrants."[8]

"Peoples" who borrow heavily find this a much more appealing sentiment than do "peoples" who lend. Although revolutionary France felt content to repudiate royalist debts, it did, as the Bolsheviks apparently chose not to acknowledge by action, ultimately consent to pay one-third of the debts. The Bolsheviks did take note of the one-third of the "sovereignty of peoples" that apparently was bound by "treaties of tyrants," but dismissed it as an act undertaken "from motives of political expediency." In this case such acts were viewed by the Bolsheviks as beneath contempt, with the overriding weight given no doubt to the economic expediency of not repaying billions of dollars in accumulated Russian debts.

The United States emerged from World War I as a net creditor with loans abroad totaling $12.9 billion more than borrowing abroad. This alone would have showed that history was accelerating: it did not need the additional touch that no sooner did the United States become a net creditor than it became almost overnight the largest repudiatee in history. Mississippi and Maryland were pikers compared to the Great Powers. Over $9 billion of the United States' foreign loans were to wartime allies and were never repaid. Since the debts were largely between governments, the losses were absorbed by the U.S. government on behalf of U.S. taxpayers. The trauma of families reading in the newspaper that they had been wiped out financially was not repeated in the United States, as it had been in France. Individual Americans emerged from World War I without severe losses at the hands of foreign borrowers and with their means to invest abroad greatly enhanced.

The United States' First Foreign Lending Boom

Neither the huge foreign loan losses of Europeans nor the failure of their own allies to repay war debts dampened the enthusiasm of American investors for lending abroad during the Roaring Twenties. Between 1919 and 1929 U.S. investments abroad rose by nearly $9 billion, two-thirds of the total during the postwar decade. By 1929 Americans owned nearly one-third of the world's total of international investments, while British holdings fell from 47 to less than 40 percent. French holdings fell even more sharply, from over 20 percent to less than 8 percent.[9]

United States investment abroad totaled about $15.4 billion by 1929, the equivalent of nearly 95 billion 1984 dollars, a figure equal to more than 100 percent of the $93 billion in U.S. bank claims on developing countries at the beginning of 1982; the nearly 10 percent average annual growth in foreign lending during the decade after 1919 compared with an average overall growth of only 2.7 percent. Unlike the foreign lending boom of the 1970s, the explosion of American foreign investment during the 1920s was directed largely toward Germany, although significant investments also flowed into Latin America. Also in contrast to the 1970s, foreign investment in the twenties resulted in holdings of foreign bonds by U.S. families and businesses instead of large foreign loans on the balance sheets of U.S. banks. Investment bankers played the role of intermediary between U.S. investors and borrowers in Germany and Latin America. Threats of repudiation or default struck directly at the financial well-being of the widely dispersed bondholders instead of threatening the solvency of a few large banks at the heart of the U.S. and world monetary system.

There are, however, remarkable similarities between the lending explosions of the 1920s and the 1970s, particularly regarding the intensity with which bankers pursued government borrowers, which by any objective standards were very bad risks. Germany, which lay in ruins after the war, saddled with massive reparation debts, in retrospect hardly seems to have been a prime prospect for investors. Yet Joseph Davis describes the New York banks' headlong rush into Germany and Latin America:

> In the absence of any previous experience and traditions, the New York banking houses plunged with reckless enthusiasm into international lending. They competed with each other and with the banking houses of other centres for every single loan transaction. It was very easy for any foreign Government, province, or municipality to raise large loans since lenders were cutting each other's throats for the privilege of satisfying the demand for capital. It was said at the time that the leading hotels in Germany had a very prosperous time because most of their rooms were taken by representatives of American financial houses who came to Germany to persuade some obscure municipality to accept a large loan. While Germany was the favourite hunting-ground, other countries also had a due share of this lending fever. Huge amounts were lent to some continental countries and to every country of Latin America. Loans were granted to provinces whose very existence was unknown until their names appeared on the prospectus. Every device of supreme salesmanship was made use of in order to place foreign bonds with an ignorant and indiscriminate investing public.[10]

Davis's description is echoed by Anthony Sampson, describing bankers nearly half a century later at the IMF's 1979 Annual Meetings:

> Many of them begin to look not so much like bankers as financial middle men, contact men, or—could it really be?—salesmen. As they pursue their prey down the escalators, up the elevators, along the upstairs corridors into the suites, they cannot conceal their anxiety to do business. For these men who look as if they might have been trained to say no from their childhood are actually trying to sell loans. "I've got good news for you," I heard one eager contact man telling a group of American bankers; "I think they'll be able to take your money."[11]

During the rush to lend abroad in the twenties there was no shortage of "sound" reasons to make such investments. Before the war Germany had been the most powerful industrial nation in the world. Americans expected that such a nation would surely recover and that the twenties represented the time to get in on the "ground floor." Most did not understand the extent of the devastation wrought by the first Great War, which, in addition to destroying most of Europe's factories, had resulted in the death of one-third of its men of the combatant generation. Nor were Americans—including their President, if one is to judge by his performance at the Versailles negotiations: ("Logic? I don't care a damn about logic. I'm going to put pensions in.")—aware of the profound residue of bitterness that manifested itself partly in massive punitive reparations imposed on Germany. Aside from short recessions in 1920, 1923, and 1926, the 1920s were an era of great prosperity. Industrial production increased by 50 percent from 1920 to 1929 (before dropping to half the 1929 level by 1932). Prices were fairly stable, and federal surpluses permitted successive tax reductions while the federal debt was actually reduced. Automobiles, movies, and radios provided delightful new ways for consumers to spend higher incomes while the ever-present chance for stock market "killings," made possible in part by margin rules geared to demented speculation, provided excitement for the smallest investors.

More cautious Americans were consoled by the knowledge that most loans to Germany were endorsed by the League of Nations—the United Nations of the twenties—much as banks are—or at least have been—consoled today by the IMF's stamp of approval. Foreign advisors were appointed to oversee the disposition of loan proceeds, and even though the United States refused to join the League, big names on loan documents persuaded the canniest bankers that the scrutiny of such prestigious "advisors" constituted a virtual guarantee of protection.

The overall pattern of the lending booms of the twenties and seventies were as similar as the intensity—during both—with which bankers sought to lend as much money as possible to economically and politically unstable countries. It was the reverse of Mark Twain's summary of the Gilded Age: "Get money, honestly if you must, dishonestly if you can." There were times when the very unappealingness of some governments seemed part of their financial charm. Intense and skillful efforts by bankers out to "sell" loans overcame any initial resistance of some borrowers who may have wondered how construction of municipal swimming baths, many of which were built in Germany during the twenties with foreign funds, would produce the means to repay the loans. Even the efforts of Dr. Hjalman Schacht, President of the Reichsbank—Germany's central bank—to dam the foreign capital flooding into the country were fruitless.

A surge of capital inflows that does not coincide with a surge in real investment opportunities can only finance more consumption, which is often concentrated on imports. This was true in Brazil, Mexico, and elsewhere during the half decade before 1982, and it was especially true in postwar Germany where, although there was relatively little damage to plants, some were confiscated and a lack of manpower and fiscal order precluded production of the radios and automobiles that were so much in demand. The flow of investment capital into Germany was very uneven during the twenties, rising sharply soon after the war and then dwindling to almost nothing in 1923 before gradually increasing again to a peak of about $2.2 billion in 1928. The second resumption of large-scale foreign lending in 1924 was followed in 1925 by a growing volume of short-term lending, partly employed to finance long-term projects. As it was to be over five decades later, the short-term borrowing was the first sign that many of the projects being undertaken with the huge inflows of capital were not "self-liquidating"—they could not generate the cash flow required to service borrowing that financed their construction, thus requiring short-term loans to pay the interest on long-term bonds. An unstable prospect indeed.

The vulnerability of an investment boom to an increase in short-term loans that must be continuously "rolled over" to keep the longer issues solvent became as evident in the late twenties as it did in the early 1980s. We have already seen how Morgan Guaranty fixed on the sharp rise in short-term debt during 1982 as narrowing the margin of borrower's and lender's security: any further shock might stop the music, leaving everyone dashing for a place to sit down while the invisible hand pulled away the chairs. International short-term lending boomed in 1928 and 1929 partly because the Wall Street boom siphoned off longer-term funds and partly because investors thought they were playing it safe by lending for six months or a year instead of for ten or twenty years.

Fueling the 1920s Boom

Another problem was that the Federal Reserve, under the leadership of Benjamin Strong, was performing an act of good citizenship that bystanders found hard to distinguish from central bank clubbiness as practiced by Strong and Montagu Norman, his British counterpart. The British had marked a very implausible return to normality in 1925 by putting the pound back on the gold standard at a dollar equivalent of $4.86—at least ten percent more than it was worth, the mark of a business and political community more preoccupied with financial supremacy and capital exports than industrial competitiveness. "The pound must look the dollar in the face" was the cry. British exports, maniacally overpriced, of course sagged. The pound's valuation became even more absurd; any decent rate of interest in New York would pull money out of London as fast as if it were Bagehot's Moon. So money had to be cheap, that is, abundant, at the New York Fed. Some of it revved the stock market up to irrational prices and some of it financed unsecured municipal improvements in the Weimar Republic. Half a century later cheap money at the New York Fed was meant to revive an economy that needed higher oil prices much more than it needed higher prices of houses, antiques, diamonds, gold, and other inflation hedges. Again, misapplied, easy money spilled abroad—this time into Latin America, Eastern Europe, and the newly industrialized countries of

Asia. Like the money that chased gold prices to over $800 per ounce, some of it was not going to come back.

The United States and its hope for ever-growing prosperity kept the twenties lending boom alive well beyond the time when it ought to have been cut back. Profits from stock market trading went to finance more short-term lending to Germany and Latin America. The short-term flows kept Germany afloat long after it had become clear that the country was suffocating under a mountain of war debts. The 1929 Young Plan to further ease reparations burdens on Germany—relaxed under the Dawes Plan—came too late. Germany had become heavily dependent upon steady injections of capital from abroad.

Lending Boom Reflects American Spirit

The artificiality of the continuation of the twenties' lending boom almost to the end of the decade is evident after the fact. But for Americans living in that first period of an earlier new economic order that saw their country taking its place as the greatest nation in the world, the future held unlimited possibilities. In 1925 they were reading F. Scott Fitzgerald's *The Great Gatsby:* "Gatsby believed in the green light, the orgiastic future that year by year recedes before us. It eluded us then, but no matter—tomorrow we will run faster, stretch out our arms farther. . . . And one fine morning—."[12] At the same time Bavarians were beginning to read *Mein Kampf*—"My Struggle"—the brooding manifesto of an obscure former corporal imprisoned after he had jumped on a beer hall platform, fired a pistol, and shouted "National revolution has broken out."

The United States reached out for Europe with capital and spirit. Charles Lindbergh landed in Paris on May 21, 1927, and the airlink between the United States and Europe was born. Just two weeks later, on June 5, Clarence Chamberlain landed in Germany after a now almost forgotten record flight of forty-three hours, ten hours and 300 miles farther than Lindbergh's. Europe reached out too with the first transatlantic flight of the German "Graf Zeppelin," which landed in New Jersey on October 15, 1928. But almost nine years later, on May 6, 1937, the German zeppelin *Hindenburg* caught fire as it was preparing to land at Lakehurst, New Jersey, and was incinerated in front of thousands of horrified spectators who had come to witness the completion of its maiden voyage. The decade of transition from the *Spirit of St. Louis* to the *Hindenburg* was a painful one that left in ashes the hopes which had carried stocks to dizzy levels and sent American money around the world. It was reminiscent of the transition from the 1970s exuberant touting of Brazil as the nation of the future to the 1983 newspaper stories of its food riots, poverty, and ruination of well-established businesses.

Collapse

Just as hopes for future prosperity were legislated into requirements by the Great Society programs in the late 1960s, so by 1929 had American stock prices reached levels that required fulfillment of extravagant hopes about the future of the economy. All that was needed to trigger a drop was some sign of a slowdown. It came soon after August of 1929. Production and income, which had been rising until the late summer, began to fall sharply. Stocks began their drop in August and fell by enough to trigger margin calls so that investors who had borrowed money to buy stocks had to put up more cash. When more and more investors were unable to raise the cash, brokers sold them out, dumping stocks in an effort to recoup money their firms had lent to their customers. Investor losses mounted, and some stocks fell so sharply that investment firms lost heavily as well and were unable to recoup the 80 or 90 percent of the stock purchase price they had advanced. By mid-October selling pressure became intense. Stock prices fell sharply on October 23 and the decline accelerated on the next day, "Black Thursday." By October 29 Standard and Poor's composite index had fallen to 162, 37 percent below its peak of 254 on September 7. Volume reached 10½ million shares, four times the average in September.

The collapse of U.S. security prices created havoc, wiping out investors and threatening the viability of the financial system. In six weeks almost 40 percent of the value of stocks, dispersed with unprecedented wideness among American families and businesses, was eliminated, creating a liquidity squeeze and evaporating the euphoric and unrealistic hopes that had propelled investors to keep buying shares. Short-term investments abroad virtually disappeared, liquidated in a mad rush to prop up the home market, and Germany, by 1929 a loan junkie, was left high and dry like an addict unable to get a desperately needed fix.

While the collapse of the 1970s LDC investment bonanza was not as visible as the stock market collapse of 1929—shares in Brazil and Mexico are not traded but rather are closely held—the losses involved in the 1982–83 were larger and concentrated in far fewer hands—the banks—than were those of the hundreds of thousands of investors stung by the earlier collapse. In a way the later crisis, concentrated as it was in one segment of the economy, might seem more manageable, but as the wounded segment was at the heart of the world's financial system, the premium on successful management of a concentrated crisis became high indeed.

The experience in Germany was also repeated in the Latin America of the late 1920's, which by 1928 had come to account for 44 percent of new direct American investment abroad, largely as purchases of government and munici-pal bonds with some private corporations stock thrown in for the heavy speculators. The attraction of the bonds was their yield: about 7 percent, or 40 percent above the return on most U.S. domestic bonds. With the exception of Venezuela, where loans were used to develop oil wells, many of the loans to Latin America were used carelessly, or worse—to retire previous debts, to prevent default on existing loans, or to cover domestic deficits, with still others used on absurd and often corrupt public works projects. Between 1924 and 1928 Colombia borrowed $150 million to construct a railway connecting two valleys separated by a mountain range 9,000 feet high. Since both valleys had their own outlets to the sea, they were already connected by low-cost sea transport. In addition, while the federal authorities were driving the very expensive railway tunnel through the mountains—later abandoned—the local authorities were building a costly road over the mountains. Needless to say, such investments did not produce the cash flow needed to repay loans; indeed they yielded a negative cash flow, so that even the money to pay the 7 percent interest had to come from elsewhere—perhaps from other loans.

With heavy borrowing for all the wrong reasons persisting until 1930, Latin American debts eventually became unmanageable. By 1935, 85 percent of Latin American dollar bonds were in default. Reports like that by H. E. Peters on Argentina had warned in advance that such issues were being used for unproductive purposes and would likely be repudiated.[13]

The Crisis Spreads to Europe

Repercussions of the American contraction reached Europe more quickly. The business slump led to heavy pressure for "temporary protection" from foreign competition—much like the relief granted American automobile manufactur-ers by the Reagan administration's ironic "voluntary" auto export restraints on Japan, which were neither voluntary nor in keeping with the free enterprise views selectively espoused by the strange, new breed of Republicans from California—and on June 17, 1930, Congress enacted the Hawley-Smoot Tariff, which levied the highest duties in American history. Exports of European manufactured goods, whose prices had already fallen, contracted. Reduced export revenues further weakened Europe's economies, which had never fully recovered from the war. Prices of the raw materials exported by Latin America fell as well in the face of less demand in both the United States and Europe. World trade contracted sharply, much as it did in 1982, when world trade dropped in volume by 2.5 percent, the sharpest drop since 1975's oil-shocked drop of 3.5 percent and a far cry from 1979's 6.5 percent rise.[14]

The 1930 plunge put heavy pressure on European banks. On May 11, 1931, Austria's largest bank, Kredit Anstalt, failed—and the shock waves spread across Europe through the close and complex ties of mutual dependence among financial institutions. Anyone owed money by Kredit Anstalt after May 11 did not get paid and in turn did not pay all he owed. By mid-June the banks in Germany were closed, and all short-term British assets in Germany were frozen. In July President Hoover managed to push through a one-year inter-government debt moratorium together with a "standstill agreement" among commercial banks not to press for repayment of short-term international credits.

But by the summer of 1931 the system was coming apart. France and the Netherlands precipitated runs on sterling, and on September 21 Britain abandoned the gold standard. Europeans, fearing similar action by the United States, converted large amounts of dollar assets into gold during the month after September 21. The American Federal Reserve responded by raising its discount rate from 2.5 to 3.5 percent, which temporarily ended the gold drain but intensified internal financial problems. Bank failures and runs on banks soared, and by January 1932, 1,860 U.S. banks with deposits totaling almost $1.5 billion had closed their doors.

The hopes of the twenties evaporated during 1931 and 1932. As if cruelly to symbolize the shift from the spirit of the twenties, Charles Lindbergh's infant son was kidnapped on March 1, 1932, never to be seen alive again. Thousands of penniless world war veterans camped in the parks, dumps, and abandoned warehouses of Washington, D.C., during that summer to beg the government for relief from three years of depression. In July they were driven out of Washington by a charge of cavalrymen led by Major George S. Patton, Jr., in an operation supervised by General Douglas MacArthur and his aide, Major Dwight D. Eisenhower. The men who were to become the nation's heroes of the next decade had to do its dirty work in the thirties.

Might One Collapse Preordain Another?

Nineteen thirty-three was a year that set in motion forces which were to affect the next half century. Adolf Hitler was appointed Chancellor of Germany on January 30. Roosevelt was inaugurated as President on March 4, proclaiming "We have nothing to fear but fear itself." There followed "the hundred days," which saw passage of the Unemployment Relief Act establishing the Civilian Conservation Corps and the establishment of federal aid to states and farmers and of the U.S. Employment Service. In the next two years, federal agencies dealing with almost every aspect of American business were created. On August 14, 1935, the Social Security Act was passed, providing both old-age pensions and unemployment compensation. This forerunner of the Great Society was based on fears of what the economic future might hold for workers and the aged rather than on hopes that the future would be better than the past. In 1958, after another world war and over a decade of prosperity, economist John Kenneth Galbraith published *The Affluent Society*, [15] which decried poverty in the public sector amid affluence in the private. The concept of aggressive increases in social programs was born, resulting in—a decade later —the Great Society, which created social programs based on hopes for a better future rather than fears drawn from a troublesome past.

Events of the 1970s revealed that the Great Society and its counterparts around the world were not sustainable without heavy increases in government debt. Attempts to cushion the impact of the debt pushed up money growth, and the resulting inflation emboldened Third World nations to charge more for their raw materials, especially oil. Hopes of the Third World for a new economic order became the fears of industrial countries, which, sensing a retreat of Western power now two and one half decades old painfully chronicled by events in India in 1947, Indochina in 1954, Suez in 1956, Algeria in 1962, and Vietnam in 1973, reached out to grab a share of control over the resources that no longer seemed as plentiful. Decade and a half long fears that industrial countries could run out of the materials they needed to ensure prosperity were rekindled with publication of *The Limits to Growth* in 1972, a year before the first oil crisis, which then struck with the added force of a prophecy fulfilled. [16]

With fortunes abroad deteriorating in the 1970s, the liberals got most of what they wanted at home. Fears of a less generous future supply of resources were not translated into a perceived need to scale down newly expanded social programs. Some doubts clouded the minds of the politicians, but they never did give up the idea that the resources for programs for which they had fought for most of their political lives could be found . . . somehow, somewhere. "Be careful what you want in your old age," said Goethe. "You will get it when you are old." It began to look as though the Western machine was growing old. The fear of raw materials scarcity became a metaphor for the general loss of self-confidence in the face of the cultural revolution of the seventies and the relative decline of power in the industrial world—like the fear of poverty among so many rich old people. The wealth was there but it just was not growing fast enough—as fast as we had once hoped it would.

Another decade passed, and in August 1982 the cycle of hope and fear turned again with the visit of an oil prince to Washington to plead for money and time. But fears for the new economic order did not mean hopes for the industrial world. A massive accumulation of Third World debt had tied together the fortunes of rich and poor countries. If anything, the message was that the new economic order had become a circle of interdependence.

The drawing of parallels between the 1920s and 1970s, together with the conclusions that might be drawn from the 1930s about the 1980s, is too crude an exercise to reveal to us the immediate future and how to deal with it. But the earlier period, with its surge of liquidity and a euphoric rush to invest in parts of the world deemed nations of the future—the United States fell into that category after the twentieth-century's first world war, hence the stock market boom, not repeated in the 1970s—does yield some useful insights into what might be transpiring two generations later. It is to be hoped that these insights can help to guide us away from disaster.

PART II

Origins of
the LDC Crisis

Chapter 3

Cultivation of Inferior Soils

> Now, as new countries advance in population, the cultivation of inferior soils must increase the cost of raising raw produce, and the division of labor reduce the expense of working it up.
>
> ROBERT TORRENS,
> *Essay on the Production of Wealth*

SINCE steam and iron made demands that could not be met from within national limits, industrial economies have been haunted by the specter of starving for want of raw materials—first grain for the growing population in England and parts of Europe, then iron ore for the industrial revolution in England, and much more recently, oil for Europe, Japan, and even the once richly endowed United States. With each step it seemed a new nightmare of deprivation loomed, threatening in ever more general terms, from classes to nations, starting in England with the Ricardian obsession to explain the distribution of income among land, labor, and capital. As a giant among English economic thinkers of the early nineteenth century David Ricardo expressed views that had a heavy influence on English thinking until well after World War I. Writing in 1919, Keynes saw the wolf back at Europe's door in 1914 in the form of ever-worsening terms of trade: "Europe's claim on the resources of the New World was becoming precarious; the law of diminishing returns was at last reasserting itself and was making it necessary year by year for Europe to offer a greater quantity of other commodities, to obtain the same amount of bread."[1] Since a part of the push behind the lending rush after the first oil crisis lay, in our view, in an acute fear of a raw resource crunch, it is useful to trace the path of thinking in industrial countries about raw material supplies since the start of the industrial revolution for some clues to forces that motivated the banks as well as the LDCs during the 1970s lending boom.

Heavy Influence of Perceived Pressure of Population on the Land

The picture in the early nineteenth century, at least according to the classical economists Robert Torrens, David Ricardo, and Thomas Malthus, was pretty gloomy unless you happened to own land. Population growth, Parson Malthus chimed in dismally, would hold wages to a subsistence level, since any wages above subsistence would just result in more offspring. It is not clear why the landed aristocracy would not be expected to breed like rabbits under this view of the world since their means were considerably above the subsistence level. Some did, of course, and in a number of notable cases produced a noticeable watering down of the stock. Perhaps Ricardo and Malthus anticipated F. Scott Fitzgerald's observation that "the rich are different from the rest of us," or the song in *The Great Gatsby*, "The Rich Get Richer and the Poor Get Children." It was the poor wage earner who, possessed of a few extra pence, was given to excessive procreation: "excessive" in the sense that bringing more children into the world meant only that they would earn a lower wage. The same fallacy of composition that produced runs on banks operated here. If only one wage earner added children to his brood while wages were above subsistence, he added to his well-being in old age since they all would earn the supra-subsistence wage and be able to provide a comfortable dotage for dear old "dad" and "mum" (provided that dad hadn't relied too heavily on the switch to keep all the "little buggers" working hard on the farm). But if all wage earners dreamed of more young ones as the way to a comfortable old age and had more children, none would obtain it since the resulting increase in the supply of labor would push wages back down to the subsistence level. Had Alfred Marshall's famous *Principles of Economics* published in 1890 been available, all this could have been said as: "the demand for labor was inelastic"; at a lower wage the total wage bill was reduced even though more people were working.[2]

As with so many of nature's rules, the balance of the population path was a delicate one with weighty implications. If, contrary to excessive procreation, laborers forewent children instead of savings—as many, especially in France, have since World War II—the result would be no one to look after them in old age. As many in the postwar industrial world have discovered, accumulated savings cannot provide companionship and guidance in old age.

Beyond the need for balance, the validity in fact of the eighteenth- and early nineteenth-century British view of income distribution is debatable. It is tempting to think that the Anglican Parson Malthus may have seen the subsistence wage as a suitable reward for a class of people apparently expected to spend any spare time copulating. Indeed neither the great Ricardo nor the giant Malthus would have fared very well with today's high priests of rational expectations, since somehow the victims of excessive procreation were expected never to learn the negative consequences of their actions and thereby modify their behavior. Of course, to be fair to the rational expectations theorists, it may be said that the working classes did not know the "relevant model" (that is, where babies came from), but surely "learning" is possible where temporal regularity (nine months) is present, though they may have been deterred by Malthus's terse—and grossly inconsistent—condemnation of contraception as "vice."

In the view of the English classical economists, the outlook for the capitalists who emerged in the nineteenth century was not much better than that for labor. According to Robert Torrens in his 1821 *Essay on the Production of Wealth,* the limitation on the supply of new land—by drainage, reclamation, or even conquest and discovery—meant diminishing returns to agriculture so that industrial countries that exported manufactured goods (for which he assumed constant returns to scale in production) would always see over time a diminution in the value of their manufactured exports in terms of imported food.[3] To put it another way, the value of what they sold (exports) would fall in terms of what they bought (imports); their terms of trade would deteriorate in line with Keynes's fears. The result would be eventual elimination of gains from trade and therefore of trade.

It seems odd for Americans in the twentieth century to think of a world in which scarcity of land and resultant shortage of food is the major obstacle to economic progress. But for a small, densely populated island nation that relied heavily on export of manufactures to feed itself, the obsession with scarcity of food is primary. It is perhaps no accident that most English food leaves visitors with the feeling that they could just as well have done without it. The British have been cultivating that feeling for centuries.

All that has been said of Britain about a scarcity of land holds as well for Japan, another small, densely populated island nation that relies heavily on export of manufactured goods for supplies of food and almost all raw materials (including oil). Like the British, the Japanese are great tea drinkers. Both come by it as great traders in southeast Asia. Both drink it between meals to dull their appetites at mealtime—almost as if hedging against the distinct possibility that the meal itself will be less than sumptuous. There are differences though. While the British present meals consistently forgettable (if one is lucky) to less-than-ravenous palates (in case the meal need be forgotten altogether), the Japanese present meals that are so pleasing to the eye that the stomach can forgive a tiny portion. As nations surrounded by water, both can do remarkable things with fish. The British plunge batter-coated bits of it into boiling oil and serve it wrapped in newsprint with fried potatoes as fish and chips. The Japanese slice it up in very artful and subtle ways and serve it on rice cakes as sushi or with soya sauce and that exquisite horseradish, nostril-clearing wasabi, as sashimi.

Fears of Shortage Generalize to Raw Materials

Fundamental to all of this is a profound concern in industrial nations about supplies of raw materials, in this case the most basic one of all—food. But the intense feelings of vulnerability extend to oil, copper, manganese, bauxite, and all of the other raw materials that great trading nations must fashion into the manufactured goods they must sell for food. This became evident after the resource scarcity scares of the early 1970s sent Japanese buyers around the world willing to pay dearly to line up reliable sources of long-term supply.

For the British and their intellectual cousins, concern about key raw material shortages persisted into the twentieth century even after the heydey of the British Empire (1870–1900), during which period England's terms of trade did not deteriorate according to Torren's forebodings, but ten years' bitter experience of U-boat blockade taught the nation how narrow its margin was. Keynes saw the Treaty of Versailles, signed in 1919 after World War I, as a tragedy for Europe (meaning for Britain). In his view the treaty disrupted a delicate economic balance that had relied on imports of food and raw materials from rural parts of Europe and overseas to supply Europe's industrial regions. Haunted by doctrines of Ricardo, Malthus, and Torrens, Keynes saw as the economic consequences of the peace an environment in which the terms of trade must inevitably go against exporters of manufactured goods.

Half a century later, in a remarkable manifestation of Keynes's dictum that "Practical men, who believe themselves to be quite exempt from any intellectual influences, are usually the slaves of some defunct economist,"[4] the demons planted in the minds of American economists by their intellectual forebears across the Atlantic popped out of President Nixon's, Ford's, and Carter's mouths as calls for "energy self-sufficiency." What Keynes failed to add was that such practical men, mouthing like smiling robots the words planted by their speechwriters, often lack even the least comprehension of what they mean. Prior to 1973, U.S. oil policy had been to avoid imports of "cheap" foreign oil lest they discourage producers at home from digging it up. Once foreign oil became four times as costly, American policy switched to encouraging (subsidized) imports of foreign oil since our own oil then seemed so precious that we ought to save it. What the American presidents of the 1970s apparently had in mind by "energy self-sufficiency" was hoarding our own oil while goading Americans, with easy money policies and underpriced oil, to guzzle up as much of the rest of the world's oil as possible, thereby leaving U.S. oil producers—who had had no little part in framing such policies—with a residual supply worth literally its weight in gold. Among the results of such absurd policies—though Democrat Carter, uncomfortable with the last implication, went for a windfall profits tax on U.S. oil companies—and their inevitable spillover to other commodities was a rush to invest in developing economies like Brazil and Mexico that were perceived to be rich in the world's diminishing resources. The aim, naturally, was to capitalize on future "shortages," which of course were not to be shortages at all but price increases.

The 1979–80 redoubling of previously quadrupled oil prices appeared to confirm this view of the world as valid for the eighties as well as the seventies —perhaps even as more valid—and money flooded into Mexico, Brazil, Venezuela, and Nigeria, then spilled over into almost every developing country, almost as if a lack of industrial infrastructure was the key to tremendous future wealth. The bubble broke when a skeptical new U.S. president—elected partially by a perception that there was indeed too much money chasing too few goods—and his banker, Paul Volcker, determined that the party had gone on far too long. In one of the least noticed yet most effective acts of the new administration, ceilings on energy prices were lifted in February of 1981, and a month later—as we shall see in some detail later on—Volcker moved in earnest to slow down the money machine.

Scarcity of raw materials was not the only bogeyman planted by Keynes in the minds of politicians. He buttressed his analysis of reasons for deterioration in industrial Europe's terms of trade with early evidence of his other great fear —that of saving. In his view the war had disclosed to all classes the possibility of consumption. For the working classes as much as for the rich the feeling emerged that if all saving could offer was an accumulation of resources to be squandered on war, there was no point in saving. The alternative was consumption. For Keynes's British mind, consumption by the working classes could only mean higher population growth, which in turn meant an "excessive population" and in turn raw materials (food) so dear that industrial goods would become too expensive to compete as exports. No exports meant no progress in a country that must, in its own phrase, "export or die."

In Keynes's view the rich were primarily of value for their propensity *not* to consume. They were to invest in an ever-growing capital "cake" that grew and grew as a monument to their nonconsumption. They were indeed "different from the rest of us." His description of saving was almost bitter: "Saving was for old age or for your children; but this was only in theory . . . the virtue of the cake was that it was never to be consumed, neither by you nor by your children after you."[5]

Most of that precious cake had been devoured by the war, along with a million of the children for whom it had nominally been saved. The rich who were left to see this grotesque mockery of half a century's accumulation for a better future were not, in Keynes's view, likely to do much saving for a repeat performance, nor did they. In Keynes's own words, when the next potlatch of civilization came, "we threw good housekeeping to the winds but we saved ourselves and we helped to save the world."[6] The sons and daughters of the Victorian aristocracy did go on a consumption binge that carried with it tragic undertones of things they wanted to forget and new directions they could not find. Of course, if this reaction meant no new cake, the capital stock would not be replaced, or at least would be replaced very slowly. As a result labor would have less machinery to work with. But similar to prophets of doom after the oil crisis more than half a century later, Keynes somehow failed to notice that the resultant scarcity of manufactured goods meant better and not worse terms of trade for industrial countries. The point is that the fear of the opposite was there, deeply ingrained in the minds of economic thinkers schooled to the high and dry perspectives of the offshore island.

Interdependence, Specialization, and Adaptability

Similarly predisposed as the island nations of Japan and Britain were to fears of raw material shortages, the actual impact of the oil crisis differed sharply on each nation. The British were cushioned by their North Sea oil, while the Japanese remained virtually 100 percent dependent on imported oil. It is no accident that Japan's collapse in growth from a 1973 annual rate of 10.2 percent to a *minus* 1.8 percent 1974 annual rate was far sharper than Britain's drop from 5.4 percent to 0.3 percent over the same period.[7] And starting with far less predisposition to lend abroad than British banks, Tokyo's big banks lent heavily to Mexico, aiming to gain a share in development of one of the world's richest suppliers of oil.

While steam and iron, like steel and shipbuilding, have made heavy demands on raw materials that had to be satisfied from without the world's great island nations, the effort has not gone unrewarded. Britain dominated the world's economy as a great manufacturing and trading nation before World War I. Japan has been the economic miracle case among the major industrial powers of the postwar period, with a phenomenal average annual growth rate during the sixties of 11.1 percent, more than double the typical 4.8 percent figure for other industrial countries. Even during the last half of the seventies, oil-starved Japan managed a 5 percent average annual growth rate against an average rate of 3.5 percent for industrial countries overall, thanks in no small part to energy conservation policies in the form of full pass-through of higher prices to domestic producers and consumers who then accelerated their adjustments to the new reality of scarcer oil. The risks of economic interdependence do have their rewards, which owe much to specialization and adaptability to changing world markets. No manager or worker at Toyota, Datsun, or Honda has ever believed that he possesses an inalienable right to keep pouring out the same product mix irrespective of world conditions—witness Honda's rapid switch to produce more motorcycles with smaller engines in the face of the absurd recent U.S. effort to shelter obsolete Harley Davidsons from Japanese competition.

One does not exaggerate by saying that the key to economic progress is specialization. As Ricardo put it, if England can produce more cloth by devoting all of its efforts to weaving while France can produce more grain, then both—and that is crucial—*both* France and England will be better off with England specializing in cloth production and France, in grain production. The reason is simple. It is the loss of cloth that follows from English efforts to produce grain or the loss of grain that results from French efforts to produce cloth that matters. "Do what you do best and trade for the rest" is the credo of comparative advantage developed by thinking of British classical econo-mists in the eighteenth and nineteenth centuries. For a small geographical area like Britain, this meant trading with other nations. The same has been true of Japan, which, coming to the game later than Britain, has been more of a doer than a thinker about specialization and trade.

The United States—both to its benefit and, especially more recently, to its harm—has been something unique among the world's major industrial na-tions. It is the only country, with the possible exception of the USSR, that has not had to rely on trade with other nations for prosperity. The reasons are its rich endowment of natural resources and large national market area. Special-ization has been regional within the United States, a nation that, by itself, had by World War I become a large enough market area to enable producers to enjoy economies of scale while being able to sell most of their goods at home.

Despite its great potential for increasing production of goods, an inevitable result of specialization limits how far it can go. Specialization leads to interdependence. The more specialized England becomes in the production of cloth, the more it must rely on France for food. For France, dependence on England for cloth increases with the quantity of resources it devotes to grain production. The essentially pragmatic political limit to the economic gains of specialization is obvious. If England relies almost entirely on France to feed its people, the French may starve England into submission at will. Classical economics to reach fulfillment predicates perpetual peace, and it is a grim doubleness of history that makes the age of greatest interdependence equally that of the greatest wars and numbers of wars. Comparative advantage bears no necessary relationship to strategic advantage, and, hard as it is to believe, even the French are more likely to put up with going unfashionable than the British will with going hungry. The British must either have alternative sources of food supply (such as a colonial empire with its sea lanes guarded by a great navy) or yield a strategic advantage to the French in order to gain an economic advantage.

It is, in view of this combination of interdependence and powerful national identities, no accident that to this day some of the bitterest disputes within the European Economic Community concern agricultural policy. Europe, including England, is a natural free trade area, interlaced with a rail and ferry system far more intricate and complete than that of the United States. Trains leaving Paris in the evening reach Munich and Vienna on the other side of Western Europe by morning. Average flying time is about an hour. But Europe also has been a natural battlefield with endless combinations of allies trying to exploit strategic advantages, few of which lack an economic component, in order to dominate the continent. The last effort by Germany involved as young men (and women: Margaret Thatcher) most of today's leaders. Someone sixty years old in 1983 was born in 1923 and was sixteen when Hitler invaded Poland in September 1939.

All of the conflicts inherent in Europe between the economic advantages of specialization and the strategic vulnerabilities apply even more powerfully for Japan. Lying in the northwest Pacific off the coasts of the USSR, Korea, and China—"a dangerous neighborhood," as Herman Kahn used to say—its supply lines are longer and its potential strategic exposure greater than, say, that of England, which could turn to North America 3,000 miles to the west, especially after twentieth-century advances in sea and air transport. By contrast, Japan lies over 8,000 miles from the population centers of the eastern United States. Even with the advent of polar routes during the 1960s, it is 6,700 miles from New York to Tokyo—fourteen hours nonstop—while New York to London is about half that and a mere three and one-half hours on the *Concorde*. Japan's history has been one of bitter military conflict with Russia, Korea, and China. Japan's alliance with Germany in World War II was tied to German aims to flank the massive Soviet Union while simultaneously, though less proximately, flanking the United States. But the actual determination to go to war was culminated by the U.S. oil embargo, which drove Japan to conquer the oilfields of the Dutch Indies. A look at the globe—especially revealing is the vastness of the Pacific relative to the effective range of 1940s vintage aircraft—suggests the madness of such a flanking scheme, but the fact that it was tried suggests also the power of strategic goals together with the desperation of a specialized industrial power faced with strangulation due to lack of oil. If Japan could be in an alliance with Germany that dominated the Soviet Union and the United States, raw materials for the industrial machines of those two countries would be available in almost unlimited supply. Again we see the power and sometimes dangerous consequences of the drive to secure raw materials by highly specialized modern industrial states faced with severe limits on land area and other resources.

The United States Learns Interdependence

Until World War II the United States was generally able to avoid this painful choice between economic and political advantage. It possessed both "grain" and "cloth"—it could meet its needs for finished goods as well as raw materials from within its own borders—and as long as regions did not attempt to overexploit the political advantages stemming from specialization, the strategic limits on specialization were less than they would be upon such regions functioning as separate nations. The major test of regional cohesion came with the Civil War. Bitter as it was, that struggle reflected how entwined the regions of the United States were after only a generation of railroads: at the height of the war, both Union and Confederate governments had to arrange for licensed trading with their adversary. A few years later, with that costly and tragic conflict behind it, the United States was able to move ahead rapidly with industrialization and attendant regional specialization. Strategic limitations upon available supplies of raw materials or existence of markets were far less telling than for other industrial economies.

"Trees," goes an old financial adage, "simply don't grow up until they reach the sky," and after World War II even the United States began to confront on a major scale the strategic limitations of a highly specialized industrial economy. During the war rubber had been a crucial concern, but success in synthesizing it begot the faith that science could find a way around any of nature's accidents. But the quantum leap in scientific sophistication and technology, especially after the *Sputnik* shock in 1957, pushed demands further. Exotic metals like titanium, manganese, cobalt, platinum, and tungsten, along with mundane but crucial tin had to be obtained from remote and not always friendly or stable sources—South Africa, the Congo, Zambia, Thailand, Malaysia, China, and the USSR. As the American economy grew, even less sophisticated materials like coal and oil, readily available at home, might be purchased more cheaply elsewhere. Another constraint emerged in the form of concerns about the by-products of intense production. Dirty smoke from coal meant more pressure on oil for energy. Always the promise was that someday unlimited energy was going to be available from almost any matter —water's minute but potent deuterium content was usually the promise of choice—through the wonders of nuclear science.

Before 1973, U.S. policy was—as we have already noted—aimed at protecting domestic oil producers from the threat of cheap foreign oil with the argument that since oil was important, you wanted to encourage companies to keep looking for it, and that wouldn't happen if they couldn't sell the stuff at prices above those available from foreign producers. The fact was that U.S. producers lacked a comparative advantage in oil production but the misuse of a strategic argument limited imports. The result was that the United States zestfully used up its own oil while foreign oil was cheap. After the first oil price shock from OPEC in 1973, the energy self-sufficiency argument produced a perverse reversal of U.S. policy that led to heavier use of more expensive foreign oil—not to mention more oil in general thanks to price ceilings—on the grounds that we had better save our own for later, perhaps until we again could forego cheap foreign oil.

Europe Learns of the United States' Vulnerability

During the interwar period of the 1920s and 1930s, Europeans surveying the wreckage of their societies had gazed wistfully across the Atlantic. There at least was an emerging industrial economy with its own resources: vast territories rich in coal, iron, cotton, corn, and timber, that could specialize, grow, and prosper without the strategic constraints that had finally brought Europe down. It took exposure to GIs in the second nasty world conflict for Europeans to recognize that the flexibility of U.S. social structures and technical competence might have as much to do with U.S. economic success as material advantage. Keynes's description of Woodrow Wilson captures well the mixture of admiration and contempt in the European view of Americans:

> The President was not a hero or a prophet; he was not even a philosopher; but a generously intentioned man, with many of the weaknesses of other human beings, and lacking that dominating intellectual equipment which would have been necessary to cope with the subtle and dangerous spellbinders whom a tremendous clash of forces and personalities had brought to the top as triumphant masters in the swift game of give and take, face to face in Council—a game of which he had no experience at all.[8]
>
> . . . this blind and deaf Don Quixote was entering a cavern where the swift and glittering blade was in the hands of the adversary.[9]

While Americans were seen by Europeans as naive, they were almost universally viewed as being, at the same time, almost unimaginably rich in things that Europe lacked. In *The American,* Henry James has a young French aristocrat address his hero:

> Being an American it was impossible you should remain what you were born, and being poor—do I understand it?—it was therefore inevitable that you should become rich. You were in a position that makes one's mouth water; you looked round you and saw a world full of things you had only to step up to and take hold of.[10]

United States performance in World War II did little to shake this view. Of course at the same time there emerged the feeling that unbridled hegemony of the United States and Americans was undesirable. They still had to live in a world with other nations whose economic and strategic concerns were to some extent legitimate U.S. concerns. And besides, Americans were naive in the ways of the world. As the world's most powerful nation, it needed European guidance.

By the 1970s hopes that the United States might buffer most resource problems for the industrial world were dealt a severe blow. The Club of Rome's 1972 report, *The Limits to Growth,* operatically articulated fears of global resource shortages, stating flatly that:

> If the present growth trends in world population, industrialization, pollution, food production and resource depletion continue unchanged, the limits to growth on this

67

planet will be reached sometime within the next one hundred years. The most probable result will be a rather sudden and uncontrollable decline in both population and industrial capacity.[11]

The Club of Rome had resurrected Malthus right down to the detail of making the ultimate disaster of unsustainable population demands on resources depend on exponential (ever-increasing) growth of the demands against only stable or even slowing growth of the supply of goods necessary to sustain life. For both doomsayers, consequences of these trends were disaster, although interestingly enough Malthus was a little more dramatic, looking to periodic checks on population to come from war or pestilence instead of the bland "rather sudden decline in industrial capacity" foreseen by the Club of Rome. The latter phrase may not have been the most effective imaginable, though its authors did not have the advantage of being close to theological doctrine as did Parson Malthus. The church knows that you cannot get sinners to change their ways without scaring them to death. Perhaps the authors of the Club of Rome report ought to have traveled up to Florence to gaze at the seven layers of Hell on the ceiling of Il Duomo. If the sixth level will not get them, then the seventh will.

"Inferior Soils"—1970s Style

Five generations—from the first steam locomotive generation to the first generation with personal computers—have passed since Ricardo, Malthus, and Torrens translated the fundamental reality of limited resources into a world where catastrophe—pestilence, famine, conquest—were part of life. Yet in 1972, after almost forty years since governments had swept downstream over the rapids of the Great Depression and the greatest war to the calm of the belief that after the fact they could always find more money, more resources to do what they needed to get done, the simple reality that some resources at least are practically limited was restated. As before, the extrapolation to disaster, in making the unwarranted passage from the grimly empiric fact that humanity can be incredibly foolish to the ill-founded belief that it is stupid, ignored two fundamental forces: the ability to substitute among resources in production and among goods in consumption—to use coal instead of oil power or fertilizer in place of gasoline and more land; to eat fish and poultry instead of meat—and the possibility of technological change. Food production is now capital intensive, not land and labor intensive as it was when Malthus and Torrens looked out over the peasantry flailing the corn in their smocks. Use of alloys and electronic fuel monitoring has resulted by 1984 in 3,000-pound cars that can do everything a 4,500-pound car could do in 1970 and get twice the mileage while doing it.

But substitution and technological change take time, and progress by way of technology is very unpredictable. When thoughtful men, fresh from a reading of *The Limits to Growth,* saw the price of oil quadruple between November 1973 and March 1974, they were galvanized by fear. Economists who argued that relative prices would adjust or that the crescendo of energy prices was a tremendous spur to technological substitution were brushed aside by people who had to line up for gasoline that cost twice what it had a few months earlier. And nobody washed your windshield or gave away maps or juice glasses. Instead gas stations stayed open for only a few hours a day and made it pretty clear that they wished you would not come around so often. It did not take much imagination to envision lining up for other things, like food; the crackpots foretelling doom and recommending essential supplies to be hidden belowground started to come out of the woodwork. For doubters there was the retort that it had all been foretold a year or so earlier. The authors of *The Limits to Growth* were saved the trouble of conjuring up vivid images of Armageddon, which, by the way, is less than forty miles from the Golan Heights, easily within earshot of the oil-financed artillery of the Arab League. It was here at gas stations a lot sooner than we thought it would be.

The first oil crisis dispelled any lingering doubts that might have remained about the shared vulnerability of all nations, including the United States, to the limitations—or constrictions—of global resources. These concerns were intensified by environmentalists who added to the hysteria by claiming that, far from looking better, the future looked bleak since ever-continued use of existing technology would cause the world to choke on its own waste before it ran out of resources.

The grim prophecy of limited resources—of "shortages" (that is, there may not be enough for *you,* my friend)—and its apparent prompt fulfillment as 1973 turned into 1974 completed the process whereby inveterate and often irrational pessimism in the 1970s replaced the jaunty, improvident optimism of the 1960s. That transition of mood had begun in 1969, a year in which on July 20 the world had witnessed the exhilarating spectacle of man's first walk on the moon and, just four months later, on November 16, had seen first reports of the shameful and barbaric My Lai massacre.

The 1970s: Hopes Unrealized

At the end of 1969, a recession began in the United States that lasted nearly a year. At the same time the Great Society programs began to come onstream with a 1970 Nixon budget that called for spending more than $200 billion for the first time and, on April 1, 1970, increased social security benefits by 15 percent. During 1971 and 1972 the red ink really began to flow in the industrial world. At first governments printed money to soak it up, with money growth averaging about 12 percent during 1971 and 1972, up from about 6 percent growth in 1969. The result was a sharp rise in the inflation rate in 1973—even before oil prices surged—to an average of 7.3 percent, more than double the 1960s average of 3.4 percent. The oil shock pushed the rate to 11.8 percent in 1974.[12]

By the middle of the decade, governments of industrial countries were pretty nearly tapped out. In Western Europe the share of government spending in GNP was averaging close to 45 percent, up from about 30 percent in 1960. In the United States the share had reached 35 percent by 1975, up from 27 percent in 1960. Government spending and deficits rose abruptly in Japan during the mid-1970s, as generous new retirement benefits came onstream. Government deficits everywhere mounted sharply in 1975 and 1976, with funds raised by governments accounting for a much larger share of total funds raised. In the United States the share rose from a 1970–73 average of 20 percent to a 1974–76 average of 33 percent; in West Germany, from 15 percent to 39 percent; in Japan, from 18 percent to 36 percent; and in Britain, from 23 percent to 41 percent. Meanwhile, output growth in industrial countries fell from an average of 4.8 percent during the 1960s to zero in 1974 and minus 1.5 percent in 1975—the sharpest recession of the postwar period. Optimistic predictions that had smoothed the way for granting more generous benefits were not fulfilled. Undaunted, President Ford requested another $1 billion for food stamps on May 5, 1975, and eight weeks later, on July 1, social security payments were boosted 8 percent in cost-of-living adjustments. Meanwhile, on June 10, the New York legislature had moved to create the Municipal Assistance Corporation to bail out holders of New York City's bonds.

Together with the pressure on governments to deliver on promises made in happier times came a feeling that the industrial world was adrift, facing evaporation of the American economic and political leadership that had characterized the postwar period. The dollar-centric Bretton Woods system had ended on August 15, 1971, though it was not finally pronounced dead until March 2, 1973, when massive sales of dollars forced a week-long closure of foreign exchange markets. The shame of the Nixon resignation in August of

1974, followed less than a year later, on April 29, 1975, by the humiliating surrender of Saigon, had taken its toll.

The mantle of hope for the future had been removed from what had seemed to many the inevitably rich United States and placed upon the countries rich in oil and rare metals. The future—for it seemed now that the U.S. frontier had disappeared, both literally and figuratively—belonged to the new economic order. Just as pessimism was, in the 1970s, replacing the optimism felt by industrial countries during the 1960s, so the reverse transposition of mood was occurring in the Third World. The men with billions of petrodollars to recycle and visions of Malthus, Ricardo, and Keynes—together with the dire warnings of the Club of Rome—in the backs of their minds took note and began to place their bets on what seemed to be sure winners.

Chapter 4

No More
Fruitless Growth

> . . . in a variety of ways nationalist motivations
> operate to make the inauguration of economic
> growth extremely inefficient, possibly to the point
> of ineffectiveness.
>
> HARRY G. JOHNSON,
> *The World Economy at the Crossroads*

A DEEP-SEATED suspicion that economic intercourse with other nations will somehow result in the loss of something precious has persisted since villages began to trade at the end of the Dark Ages. The reason for this probably lies in the fact that the gains from trading arise from specialization, which brings with it interdependence and a loss of a sense of self-sufficiency. Threats to the delicate balance among nations that arise from a high degree of interdependence result in cries for a return to self-sufficiency—recall our discussion of responses to the first oil crisis by Presidents Nixon and Ford, who put forward well-publicized plans to make the United States totally self-sufficient in energy.

We have already seen that the classical British economists of the early nineteenth century believed scarcity of land would result in a worsening of the terms of trade for industrial nations. Their fears were echoed by Keynes, who saw World War I as having wrecked the delicate balance and division of labor upon which the affluence of prewar industrial Europe had rested. Keynes foresaw the shortage of raw materials as bringing a chronic deterioration of the terms of trade of Britain and much of industrial Europe.

No More Fruitless Growth

Industrialization as the Key to Development

Thirty years later, after a worldwide depression and another world war, economists H. W. Singer and Raul Prebisch turned their attention to the developing countries of Latin America and came up with the opposite argument.[1] Those countries, they claimed, must industrialize in order to avoid chronic deterioration in their terms of trade. According to Singer and Prebisch, if Latin American countries predicated their development on supplying raw materials and foodstuffs to the industrial nations, slow growth of demand for these products —when there was no such limitation on the strong appetite in developing countries for imports from industrial countries—would steadily drive down the value of their tin, oil, copper, and manganese in terms of the desired automobiles, steel, aircraft, and consumer goods they were importing. The fruits of labor productivity increases in mining and other extractive industries would, in effect, just be given away to rich countries; and labor, if it tried to raise its wage by moving out of the export sector, would only depress wages in the domestic sector.

It took less than a third of a century for Singer and Prebisch to reverse the century-old classical view that industrialization inevitably entails a steady downward trend in the manufacturers' purchasing power vis-à-vis their suppliers. In the views of both Torrens and Keynes and of Singer and Prebisch, the culprit was raw materials. But where the classical economists dreaded the effects of scarcity upon the manufacturers, the new and very political economists decried the consequences of abundance upon their suppliers; as it turned out, both were wrong—not because their theories were faulty, but because the empiric bases on which they rested—their assumptions about growth of demand and the path of technological change—were too shortsighted.

The fact is that industrialization or the lack thereof has borne no systematic relationship to a nation's terms of trade. Britain's terms of trade failed to deteriorate during the peak of its industrial boom in the last thirty years of the nineteenth century. In 1948 the terms of trade of developed and developing countries stood at 99 and 93 respectively (with 1958 equal to 100 for both). By 1963 the indices stood at 105 and 97, indicating that the developed countries were drawing somewhat away.[2] However, from 1963 to 1972, terms of trade for both industrial countries and non-oil developing countries improved an average of 0.3 percent per year, while those for oil exporters' improved 0.5 percent per year. During 1974 oil exporting countries' terms of trade surged up by 140 percent. Not bad for raw material exporters who only a decade before had been seen as doomed to stumbling ever-slower behind the caravan of industrial growth.

Direct Investment and External Control

Notwithstanding the actual outcome, a view of the world in the 1950s that looked ahead to exploitation of the labor and raw materials of developing countries by industrial economies fit well with the underlying fears about losses from trading with foreigners and the Marxist, or just resentfully nationalist, predisposition of many economists in Latin America. This view was reinforced by the fact that many companies from industrial countries were investing in developing countries by putting up factories and digging mines. Extractive industries were well represented. The investors put up the money to install modern capital equipment and sent trained engineers and managers to oversee operations. They were clearly after the natural resources that were cheaper to extract in underdeveloped countries thanks to an ample supply of cheap, unskilled local labor and richness of natural deposits.

This rapacious quest for the earth's minerals and other natural endowments, as it came to be viewed, was most extensively pursued by U.S. firms investing in Canada and Latin America after World War II. In less emotional terms it was called direct investment. Its distinguishing characteristic was retention by the investor of control over the invested capital. U.S. firms owned the mines and oil wells and American engineers and technicians operated them, overseeing raw labor hired from the countryside in Brazil, Argentina, or Mexico or wherever the project happened to be located. This type of investment is quite different from abstract portfolio investment where the investor simply lends financial capital in parts of the world where the return offered is the highest, like British investments in the United States during the nineteenth century or, in some cases, like bank investments in LDCs during the 1970s and early 1980s.

It is easy to see how growth of direct investment in developing countries rich in natural resources and populated by an unskilled labor force heavily involved in traditional (meaning primitive) agriculture led to early fears that trade and growth meant frittering away the fruits of growth to foreigners. During 1957, twenty-three subsidiaries of U.S. and European multinational corporations were added to the ninety-seven already present in Brazil.[3] In the countryside, a Brazilian farmer might see one of these subsidiaries as the strange giant that drove a new copper mine into the landscape, dropping roads, engines, and strange-speaking people, apparently from out of the air, with whom his only contact might be an offer of work in the mines for himself or some of his sons.

At the other end of the social spectrum, well-to-do and educated Brazilians saw that modern industrial equipment and know-how was the key to tapping their own rich stores of raw materials. The speed with which technological differentials set in between nations was increasing. Until the LDCs possessed that know-how and equipment, they would have to watch most of the fruits of their natural wealth going to enrich foreigners. One of those observing this dilemma was Antonio Delfim Neto, a young professor of economics at the University of São Paulo, later to become the finance minister who presided over the "Brazilian miracle" from 1968 to 1973—years in which growth compounded at an almost incredible 11 percent a year.

But during the 1950s the "miracles" were in the future for Brazil and other developing countries in Latin America and elsewhere. The perspective of that decade was one of dependence and exploitation: dependence arising from the need to go outside to the emerging multinationals, like Exxon in the United States and Siemens in Germany, for the capital and know-how to reach and make use of one's own natural resources. The sense of exploitation stemmed logically from the condition that whatever the multinationals paid for the raw materials, the real value was in the ability to transform and process them and to move and sell them without intermediary in markets of industrial countries. The Marxist doctrine that industrial economies in their later stages must move to invest in poorer societies in order to stave off inevitable collapse united with a sense of the unretrievable—much of the investment was in mining and what it took was gone forever—to produce a sense that irreplaceable natural wealth was being plundered to feed a greedy, deteriorating (capitalist) industrialism. The rape of youth to serve the jaded appetites of an aging, decaying rentier was an easy image to conjure up in the minds of proud and thoughtful Latin Americans who evaluated their postwar experience with investment from abroad in the light of never very cheering experiences with the senior powers: the United States, overbearing in the Caribbean; the English, arrogant in Argentina; the French in Mexico. And the memories of the unsuccessful are always longer than those of people or cultures looking forward to further achievement.

The Singer-Prebisch explanation for some of the hardships experienced during the 1940s and 1950s by developing economies particularly in Latin America, with its emphasis on foreign control and exploitation by foreigners, was very appealing in countries like Argentina and Brazil, where a new generation of authoritarian military leadership was emerging amid a wave of nationalism. The message was clear. Industrialize and retain control. But that meant getting capital and technological know-how without the strings attached under traditional schemes of direct investment by foreigners. The capital could be had if the banks that controlled billions could be convinced that resource-rich developing countries like Venezuela, Brazil, Mexico, and Argentina held the key to the economic future of the world. The need was present to create the perception of a "new economic order" that placed the Third World much higher up on the economic scale.

Gaining Control in Latin America

The new order toward which Latin America began to grope in the 1960s was not entirely economic; nor was it accomplished by any singular, coherent policy.[4] The major countries there were and continue to be distinct and separate, with their own particular strengths and weaknesses. Brazil, Argentina, and Chile tended toward more authoritarian, military rule than Mexico, which came into its own largely as a result of its major oil reserves coupled with huge increases in the price of oil. It is revealing to see how the oil importers of Latin America transformed themselves into economies that attracted from the big banks in the industrial world heavy portfolio commitments that expanded even more rapidly after oil prices quadrupled in 1974.

Bankers are impressed by an aura of control, particularly where disarray has gone before. Strict military rule was imposed in Brazil in 1964; in Argentina, in 1967; and in Chile, in 1973, when the junta overthrew the Allende government. In each case the initial thrust of the new military rulers was to restore order with a strict stabilization program. Wages were frozen, government expenditure slashed, and the flow of credit to the private sector strictly controlled, all with the aim of bringing down what—inevitably, it seems—was a rampaging bout of inflation. The controls were made to stick with an elaborate set of highly authoritarian controls—you can/cannot do this or that; no "ifs," "ands," or "buts"—and strong-arm tactics against organized labor. To break the severe inflationary cycle, it was necessary to rein in sharply on promises that had been made by ambitious politicians. That was where the military regimes excelled.

Once the storm troopers had established an economic beachhead, the military regimes of Latin America set out to attract foreign investment. The primary means employed was for the military rulers to team up with highly trained technocrats who acted as brokers between them and the banks. Men like Antonio Delfim Neto, Finance Minister of Brazil between 1967 and 1974, Krieger Vasena, Argentina's Minister of Economy appointed in January 1967, and Alvaro Alsogoray, appointed Argentina's Ambassador to the United States in 1966, were all symbols of international economic orthodoxy in addition to being highly trained, competent administrators. They became "intellectual brokers between their governments and international capital; . . . symbols of the government's determination to rationalize rule primarily in terms of economic objectives."[5]

Before the oil crisis, the new military regimes of Latin America were stridently self-confident. They knew exactly where they were going: export diversification and import substitution; development of economic infrastructure including railroads, ports, and electric power; and promotion of private sector investment with the aid of capital from abroad. Previous efforts in these directions had failed, they pointed out, due to resistance of local producers who feared being displaced by larger-scale, internationally oriented producers of exportables and import substitutes. The coffee producers in Brazil that could not compete in world markets were to be gobbled up by huge combines that could. Automobile production expanded rapidly in the 1950s in Mexico, Argentina, and Brazil, but the countries relied heavily on foreign suppliers for many parts. The new regimes pressed for use of domestically supplied parts to cut reliance on foreign sources while developing industry at home. "Capital deepening," heavy investment in producer-goods industries like petrochemicals, and steel and heavy machinery, was emphasized (even in the face of opposition from local producers).

Attractions for the Bankers

For the men who ran the biggest U.S. banks in the late 1960s and early 1970s, the transformation of the major Latin American economies came as a breath of fresh air. A comfortable airplane ride from New York to Rio or Buenos Aires was like a ride through time—back to the United States of a century earlier—to a world rich in natural resources and full of men who seemed eager to get on with the job of economic development. They left behind them an ever-increasing maze of government regulations and restrictions—environmental concerns, affirmative action, antitrust—that seemed designed to frustrate at every turn the man with a better mousetrap. Their fascination was with control. Men like Delfim Neto and Krieger Vasena seemed to hold all the reins right in their hands. They set the goals, established priorities, and pulled the right levers to make things happen right away. Resistance from labor or small, inefficient local producers was simply deemed not in the "national interest" and swept aside by military decree.

Those who might have felt squeamish about the repressive policies of the "generals" were reassured by their "intellectual brokers" who buffered the shock bankers who saw themselves possessed of a "social conscience" might have felt over some of the junta's more expedient acts, particularly in the early stages of the regimes. The brokers—heads of central banks and finance ministers—were, like the bankers, really citizens of the world, often with degrees from Harvard, MIT, or Chicago.

The image of control operated in another way that was very attractive to international bankers looking toward investment in Latin America. The reins held by Delfim Neto, Krieger Vasena, and their counterparts in Venezuela, Chile, Peru, and Mexico were tied to huge government conglomerates like Pemex (petroleum) in Mexico and Nuclebras in Brazil (which was responsible for its nuclear power program). These government agencies were controlled by men the bankers could trust, and it was with these agencies that they placed their capital for subsequent dispersal to petrochemical producers, steelmakers, and modern agri-businesses that were part of the government's development plan. As has already been suggested,this arrangement opened up a tremendous new opportunity for banks. No longer—it seemed—was it necessary to invest project by project. Instead of spending months and thousands of man-hours investigating ten $10 million projects, considerably less time and effort could be expended putting together a syndicate to put up $100 million or $200 million for a single loan to a government agency. The technocrats had already—the bankers thought giddily—done their homework for them. How else could Argentina have engineered an annual expansion of manufactured exports at a rate of 169 percent from 1966 to 1970 while comparable rates in Brazil and Mexico—low only in comparison to Argentina's stupendous performance—were, respectively, 80 percent and 50 percent per annum?[6] Besides this, the investments by each government were widely diversified across industries with compound diversification possible simply by distributing loans across different countries.

The technocrats in charge of development programs in Brazil, Argentina, Chile, Mexico, Venezuela, and Peru got what they wanted too. They got independence from foreign control—or so it seemed. As long as they engineered growth sufficient to pay the service charges on their bank loans, the process of internal development on their own terms could continue. An ongoing by-product of that process—provided it proceeded rapidly enough—was the development of indigenous technology and skills to keep the development process going with ever-lessening dependence on foreign technology and skills combined with ever-increasing ability to chart their own course.

No More Fruitless Growth

Brazil Pulls Out Ahead

The great English economist Alfred Marshall once observed insightfully that "nature seldom moves in jumps." Nor does human nature and the way people operate in the economy. Some problems began to emerge to slow down Latin America's economic locomotive as early as 1970, when political pressures forced a relaxation of controls that had been instituted under the Ongania regime in Argentina. As a result Brazil began to pull sharply ahead of Argentina during the early 1970s. Under the most repressive phase of its military rule —from 1969 to 1972—after Brazil made the transition from a sharp expansion in consumer durables to one in capital goods, the latter sector expanded at almost 20 percent annually.

The bankers' intense concern with the issue of control became obvious as they looked back at Argentina, bobbing in Brazil's wake, and wondered what lay ahead. Speaking on February 29, 1972, to the British House of Lords, British financier Earl Cowley spoke with presience about Brazil:

> Not long ago, the City would not consider Brazil, but now they continually talk about the economic boom in that country. It is being acclaimed as the Latin American country of the future. . . . But the question, which has never been satisfactorily answered, is, will the economic success continue when Brazil's technocratic government relaxes its very tight control over the country and allows the whole population to take part in the running of the State?
>
> On the other hand, not long ago Argentina was considered by the City to be a country for sound investment, but now it is being held up as an example of a country suffering from economic and political upheavals. At the moment Argentina is suffering from an adverse balance of payments, inflation of 50 percent, a former dictator living in Madrid doing his best to wreck any political solution, and a president desperately trying to find an answer to what seems to be insoluble.[7]

Cowley's concern that Brazil might follow Argentina in relaxing strict government control did not deter investors from continuing to pump money into the "miracle" economy. Growth of Brazil's industrial output kept accelerating: 11.2 percent in 1971, 13.8 percent in 1972, 15.0 percent in 1973. Even the inflation rate, which had been roaring along at 90 percent a year in 1964 when the generals took over, had by 1971 dropped to 20 percent and continued to decelerate to 17 percent in 1972 and 15 percent in 1973.[8] Brazil's performance convinced the skeptics; everything seemed to be going according to plan. By 1973 its external debts stood at about $14 billion, about 22 percent of its gross national product of 63.5 billion U.S. dollars, a manageable level by most standards. Net debt was only about $7 billion, allowing for its $7 billion in reserves, or about 106 percent of Brazilian exports in 1973—a fairly healthy "debt/service" ratio compared to the danger threshold set at around 200 percent by Morgan Guaranty and positively prudent compared to the 365 percent figure to which it ballooned in 1982. In addition to its sparkling performance and modest accumulation of debt—judged by its performance and promise as viewed in 1973—Brazil's balance of payments performance looked very good in that halcyon year. Exports exceeded imports by a scant $7 million, but interest payments on the foreign debt made for a negative current account balance—exports less imports and interest payments to foreigners—of $1.68 billion, still not inappropriate for a rapidly developing economy that is importing foreign capital to finance further development.

The Brazilian economy in the early 1970s—examined closely here as the biggest of the Latin economies and in some ways typical especially after the first oil crisis—hustled along like a runner dashing downhill with the wind at his back. The capital was there when it was needed to provide new machinery, roads, and hydroelectric dams, and the export markets were growing fast, creating the need to expand. Inhaling raw materials and energy and exhaling more and more goods, Brazil took big, smooth strides that brought it closer and closer to the pack of the world's economic leaders, who were beginning to breathe harder and even to stumble—choked as they were by Vietnam and social programs that together exceeded by far the means on hand to pay for them.

Shrugging Off the First Oil Crisis: With Some Outside Help

The first oil crisis blew dirty, choking smoke into everyone's face. We know how badly the developed countries stumbled. But much of Latin America—led by Brazil with the momentum of its 1967–73 "economic miracle" and by oil-rich Mexico—only moderated its pace of growth to a rate that left it still gaining on the leading industrial nations. Brazil's growth rate dropped back to an average of 6.5 percent per annum from 1974 to 1980. Most of Latin America cruised along at between 4 and 5 percent, with a brief dip to about 3 percent during the 1975 recession. This was pretty impressive compared to developed countries, whose 1974 growth sagged badly after the oil shock: the United States, to minus 0.6 percent; Japan, to minus 1.6 percent; Germany to 0.5 percent. The year 1975 was bad too, there was some recovery in 1976–78, which tapered steadily off again to an average growth rate for industrial countries of minus 0.2 percent in 1981 and 0.2 percent in 1982. The last was to have been a recovery year.[9]

But in keeping up the pace of expansion after the first oil shocks, the Latin American economies—with heavy encouragement from bankers to the North —picked up some bad habits. The first signs lay in the balance of payments accounts. From 1973 to 1974 Brazil's import bill more than doubled, from $6.1 billion to $12.5 billion. Debt rose 37 percent, from $13.8 billion to $18.9 billion, and the debt service ratio jumped to 146 from 106 on more interest and less exports. By 1976 Brazil's debt doubled, to $28.6 billion from its 1973 level; then it doubled again to reach $57.4 billion in 1979.[10] Inflation jumped back up, reaching 29 percent in Brazil in 1974, almost double 1973's 15 percent rate. The 1974 rate for all of Latin America was about 38 percent, up from 32 percent in 1973. By 1976 it had reached 66 percent.[11]

Some Flaws Begin to Show

All of these trends continued and often accelerated throughout the 1970s. When output and prices rise together, demand is rising faster than supply. As the 1970s went on, increasing amounts of the capital flowing into Latin America went to consumption (demand) while less and less went to investment (supply). These not-so-perfect aspects of the Latin American economies began to show through more and more. Despite rising exports, the largest economies —with the exception of Mexico with its oil—really had industrialized more to produce for domestic consumption than they had to export. That meant that, unlike the developing economies of Asia whose strength came from investment expansion in export markets where worldwide competition constantly weeds out weak producers, many of Latin America's industries grew up in a somewhat sheltered environment, producing substitutes for consumption goods that had been imported a decade before. Other investments were in infrastructure capital where duplication or competition do not make much sense but messages about efficiency may be slow to emerge. It is, for example, unwise to put up two hydroelectric plants side by side, or to build capacity for the year 2000 in the 1980s, as was done in Brazil with the massive, $16 billion Itaipu Dam, the largest on earth.

Together with a conviction that the future must hold great promise, development centered on spectacular investments like Itaipu produced a rapid expansion of population, particularly in the ranks of the lower middle class, which began by the mid-1970s to wield increasing political power. In fact, growth of this new class typically exceeded growth of available jobs in industry, and so political pressure built up to employ more and more workers in government and state enterprises. As a result these rapidly expanding enterprises became less and less efficient. They began to run heavy deficits, and by the late 1970s a good deal of borrowed money was used to make up the deficits—money that did not result in any increased ability to repay—or even service—growing debt.

These currents all flowed together to require a torrent of capital inflows to keep Latin America's expansion going during the 1970s. Of course, as we have seen, the banks, flush with petrodollars, were more than willing to oblige. The machinery was in place for government agencies in Mexico, Brazil, Venezuela, Peru, Argentina, and elsewhere to absorb the huge chunks of financial capital the banks had on hand to push through the petrodollar pipeline. That pipeline carried dollars from the developed world to OPEC countries and then back—to the banks in New York and London, which then transhipped billions to the Third World, holding back 1 or 2 percent for themselves in the process. Where before the first oil shock and its validation by money creation led by the Federal Reserve, the Bank of Japan, the Deutsche Bundesbank, and the Bank of England, the Third World and especially Latin America had hummed along on normal aspiration of capital from the outside, now they were propelled forward by supercharged injections of capital. The capital was not sucked in, as it once had been, by an engine that, once fed, spewed out more and more machines and instruments of production. As the 1970s wore on toward 1980, more and more the infusions of capital from outside were blown into an engine that spewed out nothing more than the means to make up deficits in government budgets and differences between imports and exports. True, the road had turned from downhill to uphill and the wind no longer blew from behind, but it was still possible to move along at a smart clip with the supercharger turned on.

Ignoring the Symptoms

The mid-1970s was an exhilarating period for the developing countries and the banks that supplied them with capital. The strong Asian and Latin

No More Fruitless Growth

American economies at the head of the growth-rate pack had blasted through the oil crisis and managed to hit their stride at still-respectable growth rates. The banks were so impressed—and so eager to own a piece of what was new and vigorous instead of their own wheezing economies shackled further by burgeoning regulatory and resource constraints that they discarded "outdated" concepts like ratios of debt service to export earnings as a measure of ability to repay debt. Why look back at history? Look ahead! Mexico's 1977 debt service of 309 percent may look high, but when you think of what oil reserves will earn, it is nothing. Brazil's 1977 ratio is "only" 207 percent—five years ago, a disaster—but surely Brazil holds as much promise as Mexico; "Let's increase our commitment in Brazil."

And so it went. For the scientifically inclined skeptics, there were liberal doses of the patent medicine remedy of "country risk analysis"—which did no more than attach some numbers to the wildly optimistic expectations about the economic future of the Third World's strongest members. *Euromoney*—the world bankers' *Newsweek* and *Time,* appearing at a more leisurely monthly interval—published its "Country Risk League Tables" in October of 1979.[12] The tables showed Mexico and Brazil to be five-star borrowers—Canada had six stars, France, seven—needing to pay less than 1 percent over the London interbank rate available to prime borrowers. A year later, in October 1980, the same tables showed Mexico, Argentina, and Brazil among the "most improved" country risk performers. Their 1980 ratios of net debt to exports were, respectively, 205, 182, and 259.[13]

By October 1981, Mexico, Argentina, and Brazil were among the worst country risk performers. During that same year Brazil's borrowing rose by 14.5 percent, Mexico's, by 24.5 percent, and Argentina's, by 31 percent, enough to push up debt service ratios for all but Brazil, whose ratio remained virtually unchanged.[14] The only way to explain the unabated expansion of lending to these countries in 1981 is to suppose that the banks and major borrowers were looking ahead to the recovery due in 1982. Conservative bankers like Hans Friderichs, chairman of Frankfurt's Dresdner Bank—one of the world's largest, ranking right next to J. P. Morgan and Co.—saw "through the mists, rays of hope [which] point to a possible, if modest, recovery."[15] Others were even more optimistic. The American Council of Economic Advisors foresaw, even in February of 1982—after growth had collapsed to a minus 5.3 percent annual rate in the fourth quarter of 1981 and continued at a minus 5.1 percent rate during the first quarter of 1982— "brisk" growth in output "in excess of a 5 percent annual rate" for the last half of the year.[16]

Left at the Station

Needless to say, the mists shrouding the economic future parted to reveal gloom for most of 1982. Output fell in the United States, Canada, France, and Germany, with only modest growth rates in Japan and France and a virtually stagnant U.K. Notwithstanding the forced optimism of the U.S. Council of Economic Advisors—whose prognostications were clung to even by those who did not really believe them—the U.S. economy's output fell at an annual rate of 2.5 percent during the fourth quarter of 1982. For developing countries—especially oil exporters—declines of the previous two years accelerated, and with their sagging fortunes came the sickening realization that even the "cautious" seers had been wrong about recovery. The alarm bells began to go off in the engine room—the United States—in the spring of 1982, and after a few month's hesitation the Fed began to respond, shoveling money into the boiler at a 15 percent annual rate between August 1982 and March 1983 after having nearly doused the flame with money growth only a third as high during the first half of 1982.[17]

But the locomotive left the station too late to pull the developing countries at the end of the train out of the path of the debt-service cannonball express hurtling up behind it. In August the crushing debt-service and rollover burden of a mountain of debt topped with short-term loans that had to be renewed every six months crushed the boiler plate surrounding the cherished hopes of bankers that recovery would come soon enough to enable Latin America to pay the $30 billion or so of interest due on its 1982 debt of $209 billion.

Brazilians will tell you that theirs is "the country of the future—and always will be." Such jaunty self-depreciation was kind of fun for those living in an economy where—the real "miracle"—GNP growth averaged 6 percent a year over the half century after World War I, jumped to the "miracle" level of 11 percent from 1967 to 1973, and managed to continue at 6.5 percent per annum from 1974 to 1980—almost as if just returning to its normal stride after a burst of speed that had, by 1973, with a GNP of $63.5 billion, propelled it closer to the ranks of developed economies.[18]

Brazil cheerfully exploited the perception—firmly implanted in the minds of bankers with billions of petrodollars to recycle—that had emerged in the early seventies and saw it as the southern analogue of the United States of a century earlier. The rest of Latin America—its attractiveness as a vigorous area rich in resources enhanced by oil, first in Venezuela and later in Mexico —rode along on Brazil's broad coattails. The technocrats in Brazil and Argentina found the formula—a rich mixture of skill, charm, and vision—needed to project the aura that their economies were—in addition to being richly endowed with ever-more-precious oil, manganese, tin, copper, and coffee—remarkably "well managed." And of course the bankers from New York, London, Tokyo, and Frankfurt revolutionized bank practice in response to the pull of apparently rich, well-managed economies and in response to the push of petrodollars to recycle. Like a poker player holding a full house who reluctantly places a first bet lest he scare away a really big pot, the bankers spoke —warily at first—of the problems that could arise from massive recycling required after the first oil crisis. Encouraged by their governments, they undertook what outwardly they described as their "responsibility" to lend billions in Latin America, Eastern Europe, and Asia while actually chasing the finance ministers of LDCs around the lobby of the Sheraton Park hoping that those ministers would—on behalf of their governments—"be able to take their money."

As a condition for letting them "take their money," the banks had to give over control of its disposition to the LDCs' finance and development ministers. The ministers got the control they had wanted since the days when direct investment and foreign control had seemed to promise fruitless growth. With the control came the need to provide technology and skills required to make the new enterprises run, and these were not always available at home. As owners of the Three Mile Island facility learned in 1979 and holders of Washington Public Power Supply System (WPSS) bonds discovered in 1983, it takes a lot of planning and maintenance to build, operate, and earn back the investment on nuclear power facilities. Somehow investors in Brazil's huge Nuclebras nuclear power facility must have thought—if they thought about it at all —that Brazilians could avoid problems plaguing nuclear power facilities in the advanced countries. The Brazilians were not about to disabuse them, since going outside for technical help would have dispelled the sense of control they were after.

As problems began to accumulate—first with specific projects and then with the whole economy—as the world economy sagged, some extemporization became necessary. Getting to 1982 meant first abandoning debt-service ratios as a guide to sound lending when by 1976 they began to top 200 percent for the best borrowers, like Brazil and Mexico; anything that told you to stop lending to those winners had to be wrong. It was a little like 1968, when the United States Congress—faced with a U.S. money supply about to exceed the requirement of 25 percent gold backing—simply repealed the requirement limiting the cash portion of American money to four times the Treasury's gold holdings. Country risk analysis was invented to tell the bankers what they wanted to hear—what staff member of a big commercial bank heavily involved in LDC lending was going to come up with an index of creditworthiness that made his boss look like a fool?

By the second half of 1981, when even country risk analysis began to downgrade the bankers' favorites there developed, on the surface at least, a strong disposition to emphasize the view that loans to developing countries were based on their prospect for the future and not on a "temporary" slackening of performance tied to a worldwide economic slump that was due to end soon. But if the future, like Gatsby's green light, "year by year recedes before us," the terms for repayment of eight-year loans do not. We can put off realization of our hopes but the rub comes when yesterday's unrealized hopes become today's contracts that call for nearly $40 billion of payments to the banks in 1983.

Beneath the surface there began to emerge the usual paralysis in the face of a looming crisis rationalized by fear of triggering a panic, established (reassuring) positions of (responsible) government officials trying to avoid panic, and fear of alienating the subject of the crisis—if we press too hard for repayment we will get revolution and default. In varying degrees, the paralysis persisted —and even became more acute—after the revelations by the Mexicans in August of 1982.

Coming Full Circle

By 1983 the circle was closing. Writing an article entitled "Latin American Debt: Act Two" in the fall 1983 issue of *Foreign Affairs,* Pedro-Pablo Kuczynski, president of First Boston International and formerly a minister of Energy and Mines in Peru, proposed that among the important routes out of the debt crisis was encouragement of direct, private investment. Kuczynski praised Mexico for beginning to take measures to "reduce in practice the requirements for domestic ownership."[19] Restrictions in some Latin American countries on foreign, direct, industrial investment were criticized by Kuczynski as "ill-conceived." The time had come—at least in the mind of one intimately involved in both sides of the Latin debt crisis—to think of relinquishing some of the control that had seemed so precious to the countries struggling—with the warnings of Singer and Prebisch fresh in their minds—to hold on to the fruits of their growth. The huge bundle loans that the banks had made to government agencies in Latin America and other developing areas—the key to undertaking with foreign capital the huge projects like Itaipu while still retaining control over the process—were beginning to choke the Latin engine of growth. Kuczynski was urging some consideration of direct private investment to bring in capital, and perhaps a little more technology and know-how to keep it running—maybe on fuel that was a little less exotic and maybe at a somewhat slower rate, but at least still running. The circle was closing—bringing the developing Latin countries back closer to some dependency on the advanced countries—and it was not clear that they had much choice but to let it close, at least for a time.

PART III

Origins of
the Global Crisis

Chapter 5

First Things First

Our international relations, though vastly important, are in point of time and necessity secondary
to the establishment of a sound national economy. I favor a practical policy of putting first
things first.

FRANKLIN D. ROOSEVELT, speaking after the
London World Economic Conference, June 1933

I don't give a shit about the lira.

RICHARD M. NIXON,
speaking on the Watergate Tapes, 1971

ESPECIALLY in the United States, the problems of the international economy have always been considered "serious but less pressing" than domestic
economic problems by all but the usual minority of economists, exporters,
importers, and financiers. It is easy to understand that governments so see
these priorities, but it has become less and less appropriate as world economic
interdependence has grown, particularly over the last quarter century. This
result of failing to move international economic issues up the agenda as rapidly
as their growing importance over the last two and one-half decades merits has
resulted in a tendency to deal with such matters in an atmosphere of crisis
laced with reference to mysterious forces at work.

Episodic and somewhat perfunctory attention to international economic
problems may be a result of the threat of nuclear holocaust. The pressing need
to minimize the possibility of nuclear calamity has understandably absorbed
so much energy of national leaders that little is left over to deal with economics
unless alarm bells are ringing that signal a serious threat to the political
cohesion so anxiously sought. Whatever the reason, the world has lurched
from one crisis to another in the international financial arena since the end of
the 1950s, when Europeans began to question the viability of an international
monetary system that enabled Americans to buy as many goods and securities

as they wanted from the rest of the world, while the centrality of the dollar enabled the United States to drain out of the rest of the world sufficient currency to make good any such purchases in excess of sales by the underwriting of issues of U.S. government securities overseas.

Gold Price and Floating: Two Decades of Uneasy Neglect

Just over two decades have elapsed since the start of the ebullient 1960s. During the first of those decades, concern in the international financial system about the ill-defined arrangements that were meant to foster growth of world trade was episodically centered on maintenance of the price of gold at $35 per ounce—the bellweather of the Bretton Woods system. After that system collapsed in August of 1971, bankers and economists were confronted with the writhings of unpegged currencies, and since August of 1982 concern has shifted to debts of developing countries.

It is important to understand the way in which the earlier international financial "crises" stemming from gold price and currency fluctuations developed and were dealt with in order to comprehend the onset and evolution of the debt crisis of the 1980s. The political leaders of the world—and particularly of the United States—have developed standard ways of dealing with problems that fall under the heading of "international money crisis." Since their attention is usually on other things, from which they turn, often wearily, while skeptical of the motives of their counterparts abroad and the value of the advice they receive at home, old habits die hard. Problems are slow to be addressed since the people who talk directly to presidents and prime ministers —often themselves not experts in matters of international finance—rely on staff people to keep track of what is going on. And men whose words reach their leaders' ears directly want to be sure that "false alarms" do not tarnish the standing of their future counsel.

As we have already seen, many of the events that, since World War II, have been brought to the attention of world leaders by men who know they will not be viewed as time-wasters have transpired in August—a time when governments plan to rest, thereby affording an opportunity for governments that have to get something done to act: the bombing of Hiroshima and Nagasaki in 1945, the Soviet Union's hydrogen bomb in 1953, and the erection of the Berlin Wall in 1961 all came in August.

First Things First

The impact of each of these three events in the United States was enhanced by the fact that none coincided with—in fact, each followed by exactly one year —the quadrennial summer madness that surrounds the selection of American presidential candidates and the start of their final head-to-head battle for the highest office. In August of 1960 John F. Kennedy and Richard M. Nixon were squaring off for what promised to be a real thriller, especially after eight tepid years of Dwight D. Eisenhower and his two lopsided victories over the slightly out-of-focus Adlai Stevenson. Senator Kennedy, upon being nominated on July 15 as the democratic standard-bearer, had given his "New Frontier" speech, which had snapped into focus a kind of uneasy malaise Americans had come to feel after eight years of being managed instead of being led. As he had shown a year earlier on July 27, 1959, in his televised "kitchen debate" with the pugnacious Russian leader Nikita S. Khrushchev, Nixon was no pushover either.

The Gold Bubble of 1960

During the hectic early phases of their campaigns, neither candidate noticed that across the Atlantic in London, the center of the market for the world's privately owned gold, the price of gold—officially pegged to $35 per ounce by the U.S. Treasury—had begun to creep upward in August. By September 26, when candidates Kennedy and Nixon were staging the first televised debate between presidential hopefuls—launching the era when the ability to project on camera has more to do with becoming president than the ability personally to dominate other men one on one—the London price of gold had pushed above $35.20 per ounce, the level at which it pays to ship gold to London from New York.

Something was in the wind in London, and it was very closely tied to what was going on in the U.S. presidential election. John F. Kennedy proved to be a very effective campaigner and, as millions of Americans discovered on September 26, a far more effective television presence than the black-jowled Richard M. Nixon. The gold market sensed a Democratic administration might be coming. While Europeans have often expressed—with some exasperation—an inability to understand U.S. politics, they have always suspected that Democrats were partisans of soft money. Such fears catalyzed existing concerns about the dollar's link to gold. U.S. balance-of-payments deficits had dumped

$3.4 billion into foreign hands during 1958, another $3.0 billion in 1959, and were on the way to dumping yet another $3.9 billion in 1960. By the fall of 1959 dollars held abroad, largely as interest-bearing Treasury bills and notes, equaled the fiat value of the U.S. gold stock—about $19.0 billion.[1] Robert Triffin, a Yale economist, warned in *Gold and the Dollar Crisis* that since the United States was obliged under the terms of the Bretton Woods system to guarantee the $35-per-ounce price of gold—to supply gold to foreign central banks at that rate on demand—a loss of confidence in that price could create a run on the United States' gold stock, which would wreck the international financial system.[2] And, Triffin added, the more prices rose, the less goods—other than gold—$35 would buy: so as Gresham's Law predicts, the supply of gold would begin to dry up while cheap money (dollars) was driving out the dear (gold).

These concerns were all prevalent in the fall of 1960, creating a kind of hair-trigger sensitivity to the issue of the gold price and the U.S. balance of payments. On October 12 Senator Kennedy, in a speech at New York's Biltmore Hotel before the Associated Business Publications Conference, trod on ice that was thinner than he or his advisors—both with eyes riveted on the United States—suspected: "I do not minimize the importance of the outflow of gold, especially in the short run. And I would never want us in the position of being forced to tinker with the dollar in order to maintain our competitive position in the world export market," said Kennedy.[3]

These innocent-sounding views filtered across the Atlantic as a suggestion that a new Democratic President—remember FDR in 1934—might devalue the dollar against gold. Eight days later, on October 20—after the embers that had begun to glow on October 12 had been fanned to life with some help from Nixon forces—the price of gold in London jumped from $35.60 to $41.00 per ounce, a rise of 15 percent in one day by a price that until eight weeks before had been maintained constant since late 1953, the point at which the Eisenhower peace began to be believable. There were, among the buy orders coming from Europe—Americans could not legally buy directly—heavy purchases on American accounts from New York, Chicago, the West Coast, and Texas. The rich and conservative Americans who remembered what FDR had done in 1934 were not going to get caught short again. The next day the *London Daily Telegraph,* in its article about the gold price surge, said: "It was reported that Mr. Kennedy has privately hinted that if he became president in the next month's elections he might revalue gold," meaning he might raise the dollar price of gold, thereby devaluing the dollar in hopes of improving the U.S. balance of payments.[4] Senator Kennedy's press secretary, Pierre Salinger, denounced the *Daily Telegraph* reports as erroneous, saying, "Senator Kennedy has never made such a statement to anyone."[5] This was a curiously restrictive denial, which did not exclude the possibility that he had studied this contingency or been advised so to act.

The message was clear. After eight years of Republican leadership, there were fears in Europe and the United States that a Democratic victory would mean a move toward inflationary policies or expedient policies such as dollar devaluation. Kennedy's speech, carefully worded as it was, was taken as a hint that this was exactly what would happen.

The efforts by Nixon to set Kennedy up on the gold price issue are revealing. By October, the Democrat was gaining ground very rapidly and Nixon—chastened by his poor performance in the debates—was beginning to run scared. Kennedy was very well staffed and extensively briefed on most of the issues. But like most American politicians, who have to practice not to say "revaluation" when they mean the opposite, "devaluation," Kennedy and his people were not well up on international economics. The gold issue seemed an excellent way to grab some headlines that would reinforce the image of democrats as soft-money types. The whole affair did a great deal to accelerate Kennedy's education on international economic issues, and he attached higher priority to them—witness the Kennedy round of tariff negotiations and initiation of foreign exchange intervention in February 1963—than did his successor, the ever-political Lyndon Johnson.

The Message in the "Gold Bubble"

There was a good deal more to the "gold bubble" episode than a commentary on American politics. For the first time in the postwar era the world economy had talked back to the United States on an issue of economic policy. The Bretton Woods system was coming to operate in a way that made Europeans very uneasy. Americans bought and lent abroad as much as they wished. Anything not made up by sales of goods abroad or borrowing from abroad just ended up as dollars held abroad. Since the dollars were "as good as gold"—it is as fair to say that the postwar world was as much on a dollar standard as on a gold exchange standard—they could be held as reserves; and besides, they earned interest while gold did not. If private foreigners did not want dollars, they could just turn them over to the central bank for local currency—European currency convertibility was restored in 1958. The central bank in turn could hold dollars or gold. In the late fifties more and more foreign central banks opted for gold, but very discreetly by later standards, running down the U.S. gold stock from $22 billion in 1956 to $19 billion in 1960.

The dollar price of gold was a symbol of the soundness of the dollar—its purchasing power over goods, its constancy as a security. Who in the fifties would hold the ceaselessly imperiled pound, the apparently never-to-cease-falling franc? The gold bubble of October 1960 was the first sign of the dollar's displacement from absolute world monetary hegemony in the postwar era. The dollar standard was on trial, and maintaining it at an equivalence that had the emotional authority of age rather than solid rational basis was a problem that the new president must address promptly. More fundamentally, it was clear as it had not been since the 1920s, when the United States bowed to Britain in international money matters, that international repercussions of U.S. economic policies—even those tied only to domestic goals—could no longer safely be ignored. U.S. policy makers took note and set about formulating policies to treat symptoms rather than causes. They worked to avert a crisis for a decade, until 1971 when President Nixon looked again at the symptom, found it too uncomfortable, and changed the disease by cutting the link between the dollar and gold. But the real problem—a loss of dollar purchasing power—did not go away, and just over two years later the dollar was to be devalued sharply against oil, often dubbed black gold—this time not at any U.S. initiative but at the initiative of men with a contempt for paper money and that impatience for substitutes that marks those who sincerely want to be rich.

Election Winner Begins to Lose the Gold Battle

But in the fall of 1960, black gold was still years away. As has been noted, candidate Nixon did his best to turn the gold bubble to his advantage, decrying in an October 20 speech to business economists those "who loudly demand very easy credit and artificially low interest rates at all times and under all circumstances" as "practically inviting foreign banks and investors to pull out the billions of dollars that they now hold here on deposit or in short term paper." Taking more careful aim, Mr. Nixon added: "If such a sequence of events should ever develop as a result of cheap money dogmatists come to high public office, a totally stupid and unnecessary gold crisis would be brought on which could have disastrous consequences."[6] "Try as we may," the editors of the *Wall Street Journal* chimed in on October 24, "we can't see what Mr. Kennedy is proposing if not inflationary cheap money." They much preferred Mr. Nixon's "plain words": "A prime requisite of the ability to govern is the ability of modern governments to keep their money straight."[7]

First Things First

Americans were impressed by these sentiments—but not by quite enough to elect Mr. Nixon. On November 8, just 14,000 more of the 68.3 million casting ballots were counted for Kennedy than for Nixon. Especially in the heart of the Midwest, where international finance in 1960 seemed as remote as the idea of buying a car made in Japan, they liked the idea of Kennedy's "New Frontier" better than the idea of shackling American economic policies to wishes of the "gnomes of Zurich" and other gold bugs. The United States, despite the October 4, 1957 shock from *Sputnik* and the loss of 10 percent of its auto market largely to German imports in 1959, was still number one. Japan and Germany were beaten countries, everyone knew that France and Britain were in decline, and there was no trade deficit with the Soviet Union.

Besides, after the October bubble and some reassuring words from Treasury Secretary Robert Anderson, the gold price settled back down to the $35-per-ounce level. The new Kennedy administration set to work with the outgoing Eisenhower team, and on November 16 emergency measures to correct the growing balance-of-payments deficit were announced. Shortly after his inauguration President Kennedy renewed the pledge to keep gold at $35 per ounce. In March 1961 the Germans and Dutch revalued their currencies upward by 5 percent against the dollar to relieve further the pressure on the American balance of payments. A few people noticed at the time that this was equivalent to a worldwide 1 percent devaluation of the dollar, but even fewer seem to have taken this as a portent.

But the gold problem hung on. During the Berlin Wall crisis in August of 1961, the price jumped up again. In what perhaps was not coincidentally the last almost wholehearted European cooperative effort, a "gold pool" was formed in November. All the central banks pledged jointly to feed the gold market a virtually unlimited supply at the official price. In practice, the sole source of replenishment of drains on the pool was the U.S. Treasury. Under the new arrangement, gold bubbles now entailed a movement of gold from the United States instead of a blip upward of its dollar price. Gold kept draining eastward out of Fort Knox. The collapse of New York stocks in May of 1962 and the Cuban missile crisis in October both spilled gold into the vaults of Swiss banks, the strongrooms of Argentinian or Italian millionaires, and, even more noticeably, the Beirut offices of the agents of rich Arabs. The response was to pay foreigners more interest on their dollars—taking a page from the book of the "Old Lady of Threadneedle Street." In addition, American capital outflows into overseas investment—dollars that moved out of the United States —were subject to a set of constraints: the interest equalization tax in July 1963, which was strengthened in September 1964, and "voluntary" controls on capital exports in February 1965 (made mandatory in January 1968).

The leak through the gold pool poured on, although more slowly. The bigger problem was that the dollars kept moving out all through the 1960s, paralleling the chronic migration of sterling from Britain. After 1965 the U.S. inflation rate began to creep up. It had stayed below 2 percent since 1957, but in 1966 it hit 3.4 percent, the highest level since the Korean War. The problem was worse in Britain, which had to devalue sterling by 14.3 percent in November of 1967—the beginning of the end of the Bretton Woods system.[8] Britain and its industrial partners had invested a good deal of prestige and foreign exchange in the attempt to hold sterling at $2.80. After that rate had been abandoned, the floodgates burst open, as much from the psychological effect as from any practical consequence. If sterling, the currency of a by-now secondary power, could not be maintained, what could hold the obviously waterlogged exchange rate of the world's premier currency? The gold pool was disbanded in March of 1968—official losses feeding the pool had become too large. A "two-tier" gold market emerged in which central banks and other official institutions continued to pretend that the price of an ounce of gold was $35 while in the marketplace nervous investors paid up to $40 per ounce.

Dollar Standard

The end of the gold pool and the two-tier system meant that the world was now unequivocally on a dollar standard since currencies were openly pegged to the dollar irrespective of its gold value. The unofficial world was quietly accustoming itself to less pious axioms that proved useful three years and more later. No one was very comfortable with this arrangement, least of all the strong-currency Germans, who again revalued the deutschemark in October 1969.

By the end of the decade, the United States had come a long way from the 1960 election whose gold bubble had been quickly forgotten in the rush of the "First Hundred Days" of the New Frontier and the "Camelot" of the Kennedy administration, which turned American eyes toward the moon. On July 20, 1969, just one day after Mary Jo Kopechne perished in a car driven by the lone surviving Kennedy brother, two American astronauts walked upon the moon's Sea of Tranquility. The decade that had forged ahead on hope for the future had reached its extravagant goal, to put a man on the moon, with no great amount of thought given to just why it was being done and how it was going to pay for itself. All of this along with Vietnam began to take its toll. Inflation reached 6.1 percent in 1969—its highest level since World War II.[9] Americans were beginning to feel the weight of Vietnam on their spirits and their pocket-books—revelation of the March 16, 1968, My Lai Massacre was still to come that November 16, while wage increases were starting to fall behind inflation. The federal government's $3.2 billion surplus that year was the last that Americans have seen to date. By the mid-1970s Vietnam, social programs, and the politicians' imperative to postpone as much of their cost as possible to an inevitably more prosperous future brought in the first $50 billion deficit since World War II.

Dollar Standard Collapse

The year 1970 was the calm before the storm in which the dollar-centric Bretton Woods system of fixed exchange rates was finally destroyed. And once the tremor of that destruction had shaken the foundations of the postwar equivalent of hard currency money—once the fetters of "sound money" poli-cies in Germany and Japan had been removed from monetary policy of the United States—the way was open for a major break in the exchange rate between dollars and commodities, including, of course, oil.

The 1969–70 recession bottomed out in November, with the drop in output exacerbated by a General Motors strike. The previous February, Arthur Burns had been appointed the new Chairman of the Federal Reserve, succeeding William McChesney Martin, Jr., who had been appointed or renewed through four administrations since 1951. Having inherited a slump, Chairman Burns was anxious to get the economy moving again. He applied the standard medi-cine—increasing the annual money growth rate from the 5.7 percent level between February 1970 and January 1971 to just over double that: 11.6 percent from January to July of 1971.[10]

Some dollars always spilled abroad when U.S. money growth was speeded up. But this time no one wanted to hold them. Dollars were dumped on the central banks of Europe and Japan, which were obliged by the terms of the Bretton Woods system to buy them up from their citizens and corporations. During the second quarter of 1971 they had to absorb about $26 billion, up sharply from $7.5 billion in the third quarter of 1970. They had the option of holding them or converting them into U.S. gold—and early in May 1971 it became obvious that the preferences of even the most cooperative of central bankers were shifting. The Dutch, Belgians, and French took up half a billion dollars' worth of bullion, just over a million troy pounds—a 747's weight in treasure—between May 3 and May 12. Unable to withstand the pressure from the torrent of dollars, governments of Austria, Germany, the Netherlands, and Switzerland—the "strong" currency countries of the day—concertedly closed their exchange markets. And what did the Japanese do? Who asked outside the San Francisco Fed?

Four days later, on Sunday, May 9, West German and Dutch authorities announced that they would allow their currencies to float against the dollar within some range, limited but unannounced. This was to outspeculate the speculators—but events made speculators out of people and institutions far different from the cynical plungers of myth and were not to be reversed. Floating was anathema under the fixed exchange rate system; it was against IMF rules! The Swiss and the Austrians—somewhat more conservatively, as befitted countries with huge tourist and banking industries—revalued their currencies against the dollar by 7 percent and 5.05 percent respectively. The IMF—like the crew of a hurricane-dismasted ship hauling on stays that had lost their sails—held a meeting to approve, after the fact, the Austrian revaluation. The Swiss are not IMF members and, since the Fund articles of agreement "forbade" members to float, the IMF had nothing to say about the actions of Germany and the Netherlands.

The United States Responds

The American reaction to the currency crisis was like that of a family out for a Sunday picnic who spots thunderclouds on the horizon—"someone's going to get wet," they say, expecting it to blow past *them*. Leonard Silk—upon whom readers of the *New York Times* with more important things to think about than the tedious details of commerce rely to decipher for them the mysteries of economics—felt able by May 12 to reassure his flock that the "average American will probably feel little effect from European upsets." Remarkably, Silk—whose name over the article was on that day inexplicably spelled "Sik"—described the crisis as

The monetary storm that buffeted Europe [which] has shaken the German mark and the Dutch guilder loose from their moorings, blown the Swiss franc and the Austrian schilling higher, increased the sale of coffee, aspirin and electricity in Brussels, Bonn and Washington, and ruined the telephone and Telex budgets of several newspapers.[11]

First Things First

Undoubtedly the statements about coffee, aspirin, and telexes were true enough, but what Silk—along with most Americans—missed at the time was the fact that it was the dollar which, having broken loose from its "moorings" —remember that Congress had eliminated on March 18, 1968, the gold reserve requirement of 25 percent of outstanding Federal Reserve Notes—had chosen to roar full throttle out of the smooth harbor of stable purchasing power creating a wake whose waves were banging the still-moored European and Japanese currencies hard against the dock. Serious Americans—at least those looking ahead to a European vacation during the coming summer—were warned: "Your vacation will probably cost you more—depending on where you go." But, fundamentally, it was clear that in the *Times*'s view—shared by most "well-informed" Americans—that currency disorders were something that happened somewhere else.

Official U.S. response was not much more enlightened. The *Wall Street Journal*'s Washington man, Richard Janssen, reported on May 7: "Treasury officials, in fact, are so strongly in favor of continuing fixed exchange rates at present levels that they dismiss with earthy expletives the contrary view of the President's Council of Economic Advisors."[12] This description sounds uncannily like the 1983 Treasury-Council debate over fiscal deficits, the nascent debt crisis of the world's biggest lender to LDCs. Familiar too was the resulting enhancement of the Treasury over the Council in White House eyes, which saw fuming about international economic problems—especially technicalities like exchange rates—as a good sign that the Council did not have enough really important things to think about. Secretary of the Treasury John Connally said on May 6 that the current currency problem "can't be attributed to any action of the U.S. or to a weakness of the dollar."[13] According to one European official, frustrated by Washington's stony silence, "At your Treasury every day is amateur night."[14] The apparent incongruity is not far from the mark, since by 10:00 A.M. Washington time, when most of the top Treasury men stroll into their third-floor offices, the sun is going down in Europe, six time zones farther East.

In the face of a clear sign that the world was choking on dollars, the obliging Mr. Burns at the Federal Reserve pumped another $1.3 billion into the U.S. banking system early in May lest a deterioration of market conditions—that is, lower prices of U.S. government securities and higher interest costs—might compromise the scheduled floatation of a huge new Treasury loan.

Official Washington sat almost totally silent. President Nixon did not say a word, perhaps biding his time until August, probably just not interested; the newspapers had to come up with their own explanations for the crisis. The ever-arcane eurodollar was unmasked on the front page of the *New York Times* of Monday, May 10: "If there must be a villain in the current monetary crisis, it could well be the vast, uncontrolled and somewhat mysterious entity known as the Eurodollar market," wrote John M. Lee from London.[15] Eurodollars were held guilty because they "served as the vehicle for the huge speculative rush into the West German mark and other strong European currencies"— an explanation comparable to saying the ship's company capsized it by taking to the lifeboats in the face of overwhelming evidence that it had struck a rock and was going down.

Also included on the list of villains was—everyone's favorite—the growing number of multinationals with their "reasons and resources for international currency dealing on a grand scale." The *Wall Street Journal* did allow that "It is quite possible that the Federal Reserve again has committed this offense [ignoring the various consequences of American actions on international money and commerce] in recent months by expanding the money supply too rapidly. This may well have prolonged the heavy flow of dollars abroad." But the *Journal*'s editors felt compelled to end their observations on "Another Money Crisis" with some advice: "Europe's aim should be to work more closely with the United States on problems involving the dollar, and it goes without saying that the United States should reciprocate. It would be unrealistic for the idea to spread that the United States alone could keep the monetary system in good health."[16] At that moment it was the United States that was, however unintentionally, administering the poison.

Commodities Begin to Sparkle

Amid all the editorial scapegoating, Treasury expletives and deafening silence of the White House were signs that more than fixed exchange rates were at stake in the crisis of May 1971. On page 22 of its May 6 issue, the *Wall Street Journal* reported that prices of precious metals and copper surged on news of the currency turmoil.[17] After a week had passed the *Journal* reported that "Gold Outglitters the West German Mark," as the London price rose to $40.75 per ounce.[18] At the bottom of the *Journal*'s list of cash prices for commodities, in the "miscellaneous" section along with rubber and hides, sat 92 octane gasoline and fuel oil—unchanged at prices per gallon of 13½ cents and 9¾ cents respectively. In the first major postwar flight from money the desire was to possess the age-old favorite, gold. Everyone forgot—for a time at least—that beyond its symbolic value and limited applications in dentistry, jewelry, and circuitry, the yellow metal had few uses—that it was oil that made

the world's machinery run. Besides, in 1971 foreign oil was for U.S. oil companies a glut on the market. Let too much of it in and you had a "gas war" on
your hands. The folks who lined up at the pumps to buy gasoline for 20 cents
a gallon did not seem to mind, but they just did not realize that cheap foreign
oil was really a threat to national security. Of course, in just thirty months'
time they were going to have to unlearn that lesson—line up at the pumps
again—and learn that expensive foreign oil is a bigger threat to national
security—and one's pocketbook—than cheap foreign oil.

In Bed with an Elephant

The May 1971 currency crisis passed quickly enough. The U.S. Treasury
steadfastly avoided any outward appearance of concern, while internally Undersecretary Volcker and his staff were warning their economists—myself
included—not to start "blue-skying" about what they believed was the favorite panacea of the dismal science, floating exchange rates; it was not going to
be a viable option. The Japanese disassociated themselves from the "European" currency problem and denied any effect on the yen. Likely changes
resulting from the crisis were seen to include: more possible movement of
exchange rates, tighter controls on capital flows and—somehow—on the
eurodollar market, along with the perennial favorite, "better international
coordination of monetary policies." "It's hardly as big a piece of news for
the vast majority of Americans as, say, a sharp jump in the U.S. consumer
price index," opined the *Wall Street Journal,* steadfastly looking inward and
thereby failing to see the obvious link between a weak dollar in the currency
markets and a weak dollar in the commodity markets.[19] The crisis was definitely left behind after a May 25 international banking conference of the
American Bankers Association in Munich where U.S. and European central
bankers smoothed over each other's ruffled feathers. John Connally, resplen-
dent in his dark-blue banker's suit that complemented so well his mane of
silvery hair—and the silver Mercedes 600 that he liked to park outside the
Treasury in the driveway off Executive Place that led to his private entrance
—made a speech to the bankers, pledging: "We are not going to devalue. We
are not going to change the price of gold."[20] Just to show that the Europeans
had no hard feelings, Otmar Emminger, Deputy Governor of West Germany's central bank, paraphrased Henry Wallich's old joke about the U.S. as
an elephant in bed with its European trading partners: "If you are in the

same bed with an elephant, no matter how good-natured he is, it is always uncomfortable."[21]

The United States Responds Again

As it turned out, the "elephant" wasn't as good-natured as Mr. Emminger believed. During the hot Washington summer, the highest level White House staff and John Connally, the President's economic czar, were working in the utmost secrecy, with an eye cocked toward the next year's election, to put together a dramatic U.S. initiative. The May crisis had been perceived by the President's men more as a political than an economic threat and therefore deserving of attention at the highest level. Perhaps all the talk about gold had jogged the President's memory of events of just a little over a decade earlier.

The real problem Americans faced internationally—and Connally was quick to grasp it after becoming the new Treasury Secretary in February 1971 —was how to devalue the dollar, which was the lynchpin of the Bretton Woods system. The Europeans and Japanese complained bitterly about the flood of dollars, which, in fact, represented no more than payment for their huge trade surpluses with the United States. It was the old antimercantilist story: a country with a favorable *cash* balance cannot eat, wear, or live in the foreign currency it holds above its requirements of foreign goods. On the other side of the ledger, the Japanese and European surpluses were causing a deluge of complaints to the Treasury from U.S. producers of automobiles, steel, clothing, and shoes, who were having trouble competing with foreign goods made extra cheap by an overvalued dollar. The Europeans and Japanese wanted what amounted to "having their cake and eating it too." They wanted the big trade surpluses with the United States but they did not want to use the surpluses to buy dollars as the fixed exchange rate rules of the Bretton Woods system required them to do. The cautious steps toward floating taken by the Germans and Dutch in May were an experiment to cut the inflow of dollars.

But it became clear that May's adjustments were not enough to stop the dollar's hemorrhaging. The President and his men—Connally, Burns, Volcker, and Herb Stein, soon to be Chairman of the Council of Economic Advisors —waited for the biggest summer vacation weekend in Europe and on Friday, the thirteenth of August—that day again—headed out to Camp David and the relative cool of the Maryland hills. After two days of discussion of a program largely already worked out by Connally's people at the Treasury and agreed to in principle by the President, the New Economic Policy (NEP) was announced. The name had the ambiguous distinction of having already been used for Lenin's tardy recognition of market forces plus having a catchy ring that was borrowed word for word by Brazil's Delfim Netto to describe his 1979 plan to "grow" out of inflation while indexing wages. Domestic aspects of the 1971 American version included a ninety-day freeze on wages and prices, an investment tax credit, and an end to the excise tax on automobiles. The object was to stimulate spending in the economy while keeping a lid on wages and prices. The effect—it turned out—was to foment the pressures that increased wages and prices while holding them back in the short term, all of which helped blast up prices of petroleum and everything else when the controls were lifted during 1973 and 1974.

The blockbuster parts of NEP were its international provisions: a 10 percent tax on the value of all imports; a formal end to convertibility of the dollar into gold—even for foreign governments; and the floating of the dollar: the United States would neither buy nor sell currency on the foreign exchange markets, letting exchange rates seek their own level or be set by the intervention of other governments.

Floating the dollar and cutting the link to gold were against IMF rules, though no one at Camp David bothered—or dared—to mention this to a determined President Nixon. It would only have lowered his opinion of them, relegating the offender to the dreaded category of "technician." Arthur Burns was unhappy about breaking the gold connection—as if it meant anything in the wake of the higher dollar prices of everything else with which the market was beginning to register his recent dollar printing spree—but he was going to have to be the one to deliver the news to his dour-faced cronies in Basel, where European central bankers gather at their glumly exclusive "club," the Bank for International Settlements. On Sunday evening, Burns sent off telexes or personally telephoned his closer colleagues like Otmar Emminger of the Bundesbank, not reaching some until well after midnight. Administration outsiders at State and the National Security Council—including Henry Kissinger—got the word Sunday afternoon. The President went on television Sunday evening to announce the program. The IMF's Managing Director, Pierre Paul Schweitzer—like most Alsatians, a good world citizen, kin not only to the German Albert Schweitzer but the Frenchman Jean Paul Sartre and, perhaps most surprisingly, the American Eugene V. Debs—was invited by Connally over to his Treasury office to watch the television message. Schweitzer was not overjoyed by the hospitality, which featured an American President thumbing his nose at the IMF. It was a bit like being invited by the police to watch movies of someone stealing your car without having them offer any hope of getting it back.

Burns, born a subject of the Emperor Franz Josef and living in a world that rather faded away beyond Europe and the United States, did not bother to call or telex anyone in Japan. The Japanese ministers and central bankers, no less than ordinary people, got the news on television and over the news wires. Japan has never forgotten the Nixon *shokku,* which it deeply resented at the time as a symbol of its second-class status in the eyes of Americans and Europeans.

"The Most Significant Monetary Agreement in the History of the World"

Nixon and Connally got their way. When the exchange markets opened on Monday, the dollars poured into central banks in Germany and Japan. The Japanese tried to hold the rate at 360 yen to the dollar and had to buy $5 billion in a week from traders rushing for yen before their inevitable appreciation against the U.S. currency. Overall the Europeans and Japanese absorbed almost $12 billion more in "unwanted" dollars, resisting the dollar devaluation that Connally's brutally effective plan was shoving down their throats. By December, after having been taunted by Connally with a possible 15 percent dollar devaluation—horrifying for their exports—they had had enough and signed the Smithsonian Agreement, which repegged exchange rates at levels that depreciated the dollar by an average 8.5 percent against other major currencies—enough, the Treasury mavens, crying "its our turn for a surplus," thought to give the United States a swing of about $10 billion in its balance of payments. It was not enough.

The Smithsonian Agreement was signed just before Christmas at the old Smithsonian Institution building just off the mall leading from the Washington Monument to the Capitol. The whole American initiative since August had been elaborately orchestrated, and the President stuck to the score, flying up from the sunshine of Key Biscayne to tell the press—lest they did not realize it—that they had witnessed the signing of "the most significant monetary agreement in the history of the world"[22]—quite a modest consummation, really, beside 1969's "greatest event since creation," the moon landing. The Washington correspondent for the London *Economist* began his December 25 story on the Smithsonian Agreement: "The most important point about the new pattern of world exchange rates is that it will not last for long."[23] It did not—but even the anonymous correspondent might have given more than its pathetic fourteen months.

Gold, Oil, and Debt

The decision to break the link between gold and the dollar, which had stood for over thirty-seven years—since 1934—was a calculated, expedient act conceived by two very political animals, Richard M. Nixon and John Connally, who behaved as economic men only in their private affairs. It marked the final episode in the collapse of the Bretton Woods system, which had begun to come apart a little less than four years earlier with the clumsily executed and long-delayed November 1967 devaluation of sterling. Nixon and Connally had taken a page from Machiavelli's *Prince* and their goal was one of domestic economic policy not one closely tied to the international role of the dollar. The only check on the easy money policy needed to reelect Nixon had been the link of the dollar to gold and, through that, to other currencies. With that impediment gone, the rate of money growth was raised by over 40 percent in 1972 —to 9.2 percent from 6.5 percent in 1971 and 5.2 percent in 1970. As a result the American economy boomed in 1972 and 1973, growing at real annual rates of 5.7 percent and 5.8 percent—the highest levels since the mid-1960s and well above the postwar average.[24] The spell of the sixties returned; and many people saw the fleeting and heated response to overstimulus as a return to normalcy. The consequent disillusion when this brief revival of vigorous growth ended was a potent factor contributing to fears that the industrial countries were, both literally and metaphorically, running out of fuel at the dawn of a new economic order. Those who scoffed at the Club of Rome came in for a rude shock just a year later. By 1974, the 3.4 percent inflation rate of 1971 and 1972 had reached 12.2 percent. The rate had been 8.8 percent in 1973, before the impact of the quadrupling in the price of oil hit.[25] The choice had been made to press hard for expansion of demand just before an event that brought home, as never before, the meaning of a supply shock.

In 1971 OPEC had been in existence for eleven years. Very few people had even heard of it or if they had, they saw it as a bunch of sulking Arab princes seeking unsuccessfully to tout their own importance. It is very unlikely that it played any role in the decision of Nixon and Connally to cut loose from gold. Connally, a Texan, had undoubtedly heard of OPEC, but did not consider it significant enough to describe to his President—again fearing to trouble the chief with matters not of interest to him. The issues before all-night-thinking policy makers at this time were almost exclusively national, with certain European overtones. Even Japan was a cloud no bigger than a man's hand. They did not foresee that their decision to pump up demand would contribute crucially to the head of pressure that would two years later blast the price of oil from that of a mundane commodity to that of a highly cartelized scarce good. The decision of two U.S. politicians in 1971 to finance a long war and buy time for this and other foreign initiatives by going along with a multiplication of social programs without regard to international economic responsibilities did much to enable South American politicians and Arab grandees to double the price of oil in October 1973 and, dizzy with success, two months later to double it again. This quadrupling of oil prices over less than six months produced the first oil crisis, a major consequence of which was the creation overnight of a huge new class of borrowers and lenders. The ultimate lenders were the oil exporters; but possessing little in the way of banking facilities, they turned to the big commercial banks in London and New York, which were anxiously eager to recycle the annual tens of billions of petrodollars that came to provide half their profits. The borrowers at first were the oil importers, especially the developing countries unable to finance required imports of capital goods out of current earnings. As time went by the capacity of some oil exporters to absorb resources exceeded even their tremendous earnings, and, like Mexico and Venezuela, they became heavy borrowers.

The eleven-year path from August 15, 1971, to August 13, 1982, is strewn with the dissolved dreams that were in the minds of Richard Nixon and John Connally on that steamy Sunday evening in August. They watched together from the President's mountaintop retreat at Camp David the stunned reaction of the world. Out in the mist there were new princes, waiting. Eleven years later one would emerge to close the circle with a shock wave that had been building momentum since 1971, one that, rolling from the Maryland hills, passed Europe, the oil lands, and Japan to break upon the United States from beyond the Rio Grande.

Chapter 6

Demand Does Not Create Its Own Supply

America's efforts will be directed toward maintaining the strength of the dollar . . . injecting new purchasing power into the economy through a major tax cut, reducing unemployment and bringing inflation under control.

JIMMY CARTER, January 4, 1978

THREE profound changes in the world economic environment—in addition to the Pollyanna view of economic reality espoused by President Jimmy Carter and his crew of dreamers—were tied closely to the revolution in bank lending strategy of the 1970s: exchange rate gyrations, soaring commodity prices, and stagflation—the simultaneous appearance of inflation and economic stagnation, a negative version of having your cake and eating it too. Any one of the three would have thrown governments, bankers, and businessmen of the 1960s into total confusion. Coming all at once, as they did in 1974 and 1975, they left everyone running scared. After a decade during which the worst international financial disaster imaginable had been a break in the link between gold and the American dollar, the 1970s induced an atmosphere of unreality and impending doom—or at least a sense that the world as we had come to know it since the Second Great War was behind us forever.

Inflation and Then an Oil Shock

The U.S. New Economic Policy initiative of 1971 set the tone for a decade of unprecedented economic events. The dollar—whose link to gold really had been severed three years earlier, in 1968—had to be cut loose from gold. The gap between $35 per ounce and reality yawned far too wide. The way was open for sharp changes in the dollar price of other commodities, especially those whose supply could be cartelized and controlled. By 1973, after another 10 percent devaluation of the dollar in February, followed in March by a closure and then a reopening of the foreign exchange markets with rates floating and most important a good deal of inflation—much of it understated, thanks to Nixon's wage and price controls—the attack of Israel on October 6 by Egypt and Syria emphasized militarily the coming power of Middle East oil producers. On October 16 Arab oil producers embargoed oil sales to supporters of Israel—the United States and the Netherlands—and then followed up with a 5 percent cut in production. The posted price of Middle Eastern crude oil rose from $3.30 to $5.11 per barrel. But it was soon clear that the Arabs had underestimated their power. Prices on the spot oil market shot up over $5.11, and by early 1974 the Arabs were getting an average of $14.00 a barrel for their crude oil; shades of the October 1960 gold bubble.[1]

The delighted OPEC princes were reaping what a surge of world liquidity —created to finance swelling budget deficits of the big powers—had sown during the early 1970s. With inflation repressed in the United States, shortages began to develop in anticipation of the higher prices to be had once controls were lifted. Shortages created bottlenecks, which made buyers anxious to stock up. When these pressures combined with the fear of a possible cutoff of supplies from the Middle East, oil buyers responded like heroin dealers who have just been told that the poppy crop has been wiped out. But the oil boom was part of a broader trend. Pushed along by ever rising inflation in the early 1970s, the idea got out that commodities were a better way to store purchasing power than money or financial assets. A rush ensued to get out of money and into "real" assets, and barrels of oil seemed just as good as—in fact, better than—bars of gold. The buyers who bid oil up to $11.65 a barrel remembered the warnings a year earlier of the Club of Rome, and those who moved in early, after "only" a doubling of the $3.05 price in a month's time, made a good deal of money—in sharp contrast to the experience of those who bought Treasury bills in 1973 and earned about 7 percent —just about 2 percent short of the rise in prices between Christmases in 1972 and 1973.

Demand Does Not Create Its Own Supply

Overnight: A New Class of Borrowers and Lenders

Viewed in terms of the business of banks, which is to borrow and lend—or, as we shall see in chapter 7, to lend and then borrow—the oil crisis of 1973–74 was the answer to a banker's prayer. Virtually overnight there appeared what seemed to be ideal customers for both sides of the balance sheet. Governments of oil-importing countries faced with a fourfold increase in the price of oil needed to borrow huge amounts to buy the products necessary to keep moving ahead with their production plans. This was especially true of non-oil-exporting developing countries like Argentina, Brazil, and Korea, which did not have the domestic oil reserves of a developed country, like the United States, or the ability to adapt and diversify exports, like Japan. The current account deficit of the combined non-oil-developing countries rose by over 300 percent, from $11.6 billion in 1973 to $37.0 billion in 1974 and to $46.5 billion in 1976.[2]

The only thing following the first oil price increase more remarkable than the explosion of borrowing needs of oil importers was the immense combined surplus of the oil exporters. The combined current account surpluses of oil exporting countries rose over tenfold, from $6.7 billion in 1973 to $68.3 billion in 1974. Oil exporters were able, by means of a radical increase in consumption, to reduce their combined current account surpluses to $35.4 billion in 1975, some $11 billion less than the combined deficits of non-oil-developing countries.[3]

At first most of the huge increase in oil export revenues flowed to the world's largest commercial banks. The oil exporters faced with a new flood of funds tended to be cautious at first, and some came to financial centers like New York and London requesting overnight deposits in the hundreds of millions of dollars. As time went by it became clear that the overnight money would be rolled over and that more would be coming in. Requested deposit maturities lengthened and the banks were faced with the "problem" of recycling petrodollars.

More Adjustments to Costlier Oil

The atmosphere in the wake of the first oil crisis was predictably chaotic. Currencies of the major industrial oil importers—Germany, Japan, and

France—which had been strengthening against the dollar, especially in the middle of 1973, plunged by the end of the year. As it turned out, the trumped-up but unsubstantial commitment to stable exchange rates that had come out of the ballyhooed Smithsonian Agreement had ended just in time, in March of 1973, to avoid having a currency crisis on top of an oil crisis. It was obvious that some sharp realignments of currencies conditioned on dependence upon imported oil were in order.

Japan, it was soberly predicted, was likely finished as the fastest-growing major industrial power; yet by 1976 Japan's growth rate was back up to 5.3 percent against the 5.0 percent average for industrial countries, with inflation below the average of 7.6 percent. By 1979 the growth/inflation figures for Japan were 5.2/2.6, well ahead of the average 3.4/8.0 combination for industrial countries, which in turn were an improvement on the U.S. combination of 2.8/8.6.[4] Japan's remarkable resiliency was indicative of two things that were to affect powerfully the play of events after the oil crisis. Japan grew up as a trading economy—a buyer of raw materials and a seller of manufactured goods—used to adapting to sharp changes in markets for exports and imports. Second and more closely related to the oil crisis was a longstanding tradition in Japan of conservation—of oil and all precious resources taken for granted in the United States. Japan's production machinery was remarkably space-intensive and energy-efficient in 1973. By 1976 it was even more so. To a somewhat lesser extent—especially with regard to energy conservation—the European economies also had a comparative advantage over the United States. Although it had more oil than Europe or Japan, the United States used it like a shambling bachelor uses space in a four-bedroom house. Getting used to expensive oil was very difficult for Americans—like moving our four-bedroom bachelor into a studio apartment. The adjustment is far easier for the tidy fellow who's already been living in a one-bedroom apartment.

Together with gas lines, colder houses, hurried purchases of insulation, higher heating bills, 55-mile-per-hour speed limits, and a barrage of government rhetoric about conservation, energy-independence, and President Ford's "whip inflation now" (WIN) programs—all cushioned by government programs that kept energy prices artificially low—the United States and the other industrial economies had to endure the new phenomenon of stagflation. In 1974 demand was reined in by tight money policies—U.S. money growth was only 3 percent and the world average was 7 percent between February of 1974 and February of 1975. Over the same period consumer prices rose 11 percent in the United States and an average of over 12 percent in the rest of the world —partly reflecting the latter's heavier dependence on imported oil and partly reflecting European and Japanese decisions to let higher oil prices instead of government hot air induce energy conservation.[5]

Stagflation and Overborrowing

A short economics lesson helps to understand stagflation and then to see why U.S. economic policy after the first oil crisis was so disastrously wrong. Its effect, as will be seen in this and the next chapter, was to reward what normally would be gross overborrowing by those—like LDCs—possessed of the world's precious raw materials, not to mention those who rushed to buy gold, antiques, jewelry, real estate—anything *not* money. Such rewards, the result of seriously botched U.S. economic policies, served to reinforce investors who had ignored government promises to control inflation while punishing those who had believed them. After the second oil crisis, Paul Volcker quite properly decided to return the American government to the side of the angels and reward those few who believed his pledge of a move toward stable prices. The legions lined up on the other side of the battlefield included the big banks and the heavily indebted LDCs, along with real estate dealers, antique dealers, jewelers, gold bugs, producers of big, heavy cars, short sellers of government securities, and Jimmy Carter. They were all mowed down like infantry advancing across an open field on a hidden machine gun emplacement.

Stagflation was a term invented in 1974 and widely touted as a phenomenon that had economists baffled. It really was not all that baffling. According to standard economic doctrine—described by the Phillips Curve, named after the British economist A. W. Phillips—if you tightened up on demand, either by tight money or tight fiscal policy, although you had to put up with lower growth and higher unemployment at least you got rid of some inflation. The "trade-off" between inflation and unemployment described by the Phillips Curve—never solidly based in economic theory and widely contradicted by many careful econometric studies over time and in different countries—was a very appealing concept to policy makers. After stimulation of the economy had resulted in some expansion, unpopular inflation was the "price" you had to pay. Conversely, while waiting for tight money to control inflation, higher unemployment was the "price" you had to pay. The men who ran policy—always casual about economic theory and preferring to trust their "feel" of things—saw the Phillips Curve as capturing what economics was all about:

you didn't get something for nothing. The fact that a look at Phillips's original data—based on the relation between unemployment and the rate of change of money wage rates in England between 1861 and 1957—was not particularly suggestive of a stable "trade-off" or of the contention that there was no reason in economic theory to suggest a stable link between the rate of change of prices and the percentage of the labor force unemployed failed to impress policy makers who found it a "useful" concept. Like the Laffer Curve, with its suggestion that lower tax rates could mean higher tax revenues, the Phillips Curve—based as it was on careless characterization of empirical facts—was a poor guide to policy because there were circumstances—foreseeable by any competent economist—under which it was dangerous and inappropriate. The 1974 industrial, oil-importing world was one of those circumstances.

If the Phillips Curve had a foundation, it rested on two tenets: a predominance of changes in demand over changes in supply and sticky prices—in particular wages that rise or fall more slowly than prices and interest rates, which rise or fall more slowly than changes in the inflation rate thereby causing changes in inflation to affect investors who are concerned about "real" returns instead of returns unadjusted for inflation. In effect, labor suppliers and lenders must be "fooled" by inflation for the Phillips Curve trade-off to work. If they are, then a speed-up of inflation will for a time lower "real" wages (the ratio of wages to prices) and "real" interest rates (market rates less inflation). Businesses will then hire more workers at the lower "real" wage and invest in more capital at the lower "real" interest rate, thereby stimulating the economy to higher growth and lower unemployment in return for the "price" of higher inflation. Somehow the workers and investors who get "fooled" by a speed-up of inflation are supposed never to learn—and that shaky notion is what is criticized by the rational expectations theorists, who argue for no trade-off between unemployment and predictable inflation. Quite logically the rationalists add that if governments are known to respond to higher unemployment with inflationary policies, most inflation will be predictable and therefore no one will be fooled. The upshot of the rationalist view—derived from and shared by monetarists—is that no inflation-unemployment trade-off exists that is stable enough to provide a useful guide to policy makers.

But these important and compelling ideas, which began to receive widespread attention among professional economists only with the publication in the June 1973 issue of the influential *American Economic Review* of an article by Robert Lucas[6]—the high priest of rational expectations now generally agreed to be Milton Friedman's successor as dominant intellect at the University of Chicago—were too new and experimental to guide policy makers in 1974 (although they could have gotten the gist of the warnings about the

Demand Does Not Create Its Own Supply

Phillips Curve—enough to avoid real problems—from over a decade of Milton Friedman's objections to it).

Demand and Supply

The other potentially misleading tenet behind the Phillips Curve, a predominance of changes in demand over changes in supply as the major force determining the business cycle, was totally inappropriate in the wake of the oil shock—more inappropriate, in fact, than at any time in the last century and certainly since 1936, when John Maynard Keynes in his *General Theory of Employment, Interest, and Money* had introduced the idea that an inadequacy of demand was the major bugaboo behind the depression and that in order to avoid it governments needed actively to pursue policies of demand management (active monetary and fiscal policy), especially to get out of slumps.[7]

In addressing the question of whether there was a natural market process that assured full employment of labor and capital—in the wake of a period when disastrous unemployment had raised serious questions as to whether such a process did indeed exist—Keynes was inventing modern macroeconomics, the branch of the discipline that addresses the question of whether everything produced gets bought quickly enough to avoid prolonged idleness of labor and capital. Prior to that time, Say's Law, named after the nineteenth-century French classical economist John Baptiste Say, which stated essentially that supply creates its own demand, rendered absurd the possibility of demand inadequacy. The primary, microeconomic, concern of economics was to analyze efficient allocation of scarce resources that provided as far as possible for satisfaction of "unlimited wants" out of "limited resources." The limitations on available resources were obvious especially to the classical British economists whose fears about raw materials—so influential on Keynes and later, the Club of Rome—we have already seen. To those who from time to time doubted "unlimited wants," Malthus had in Keynes's words "disclosed a Devil"—the ever-present pressure of population growth.

For almost four decades the fledgling science of macroeconomics had, under the influence of its founder, been looking over its shoulder lest the bogeyman of too little demand creep up and somehow cause one of Keynes's "slips twixt the cup and the lip" on the way to facile fulfillment of Say's Law. Between earning and spending, those who earned could hold money; under some conditions, Keynes and his early followers believed, the preference for cash could become absolute. A "liquidity trap" could result given the expectation of widespead inflation into which would pour—like blood from a severed vein—the spending power needed to keep the economy alive. Even without a hemorrhaging into the liquidity trap, an effort to pump up activity with easy money might still not work because its downward impact on interest rates might not reliably stimulate investment, which Keynes saw as heavily dependent on the "animal spirits" of businessmen.

The first oil shock caught most economists—schooled for thirty-five years to worry about too little demand—like the hapless soldier who, looking behind himself for a sniper, steps smack onto a land mine. And it hit every single economy at once with a force directly proportional to the widely divergent shares of imported oil in the production mixes of developed and developing economies. It was, of course, a major supply shock. It meant that more labor time, more machinery, and more land had to be devoted to paying for the oil needed to run the economy. As a result each economy could—under circumstances existing in 1974—produce less at every price level. As economists would say and as everyone was soon to discover, the aggregate supply schedule had shifted back to the left.

As every student knows after a few weeks in an introductory economics course, given a normal, negatively sloped demand curve, a backward shift in supply means less output at higher prices—stagflation, when you look at the situation in the aggregate.

First Responses to the Oil Shock

Notwithstanding the facility of hindsight, the "flation" part of stagflation was immediately obvious to policy makers in Washington, Bonn, Tokyo, and elsewhere early in 1974. They had just come through the inflationary early 1970s led by the easy money policies of an election-oriented U.S. president who had freed himself in 1971 from the constraints of gold. Initial momentum was for disinflationary policies, which were followed except in Japan. There, uncharacteristically, an initial attempt was made to stimulate the nation out of the supply shock with a heavy dose of demand, which only made the problem worse. Inflation jumped to over 25 percent during 1974 in Japan, over five times the average rate for the decade before the oil crisis. Elsewhere money growth was cut and policies described as "deflationary" were followed. Of course they were in truth only "less inflationary" in the face of a negative supply shock that pushed prices up rapidly in 1974 and 1975 while simultaneously—as dictated by what amounted to a confiscation by OPEC of over $60 billion in resources in 1974—output growth fell from over 6 percent during a heavily demand-stimulated 1973 to zero and negative levels in 1974 and 1975. At the same time inflation rose sharply in 1974 and 1975 to an industrial country average of over 11 percent in both years; without tight policies it could have been worse, as it was in Japan.[8]

Too much was happening at once during those two years in the middle of a decade gone helter skelter. Exchange rates were gyrating as judgments were made and remade about which countries were or were not adjusting to oil costing four times what it seemed it should. Interest rates, propelled by instantly elevated fears of inflation, rose to record levels—records at least until the early 1980s. The sharpest recession since World War II hit the industrial world and financial failures like that of the Franklin National Bank in October 1974—the largest bank failure in U.S. history—and the failure of the Herstatt Bank of Cologne in June of 1974 (which sounded to the English-speaking world's ear much like the Kredit Anstalt, a bank that failed almost four decades before), reminded those over fifty and those who knew their history of the Great Depression.

If all this was not enough to convince one that the economic system was coming apart, there were ominous cracks in the image of political cohesion—some closely tied to economics. Watergate and the first resignation-in-disgrace by a U.S. president in August of 1974, New York contemplating default on its bonds in 1975, and the surrender of Saigon to the Vietcong on April 29, 1975, gave the apocalyptically inclined plenty of food for thought.

Searching for some guidance in the turbulent early weeks of his unique accession to the U.S. presidency, Gerald Ford called a September 27, 1974, "Economic Summit" meeting attended by 800 hard-hitting economists, businessmen, and financiers—a sort of quickie brain-trust approach to the obvious problem of "what next?" Whether the summit delegates or the President or both were still mesmerized by the oil shock, the result appears to have been a conviction that in the face of an oil tax equal to the total of personal saving by Americans in 1974, inflation was the nation's number-one problem. About a week and a half later, on October 8, the President announced a "Whip Inflation Now" (WIN) program composed of WIN buttons for government employees and other interested parties and a 5 percent income tax surcharge. Ironically, that was the very day when Franklin National Bank was declared insolvent, evoking shades of the result of deflationary policies in the 1930s. October 1974 also marked the beginning of the sharpest U.S. recession in postwar history. After only three months had passed, the Ford administration perceived that recession, not inflation, was its biggest problem, and on January

15 the President proposed substantial tax reductions, reversing the thrust of WIN. By May the Republican President was requesting an additional $1 billion for the government's rapidly expanding food stamp program.

The confused U.S. reaction to the onset of stagflation was fairly typical, although U.S. policy initially lurched more toward restriction immediately after the oil shock than did policy in Germany and Japan. The result was that by September of 1975, the U.S. dollar had almost recovered, back up to its Smithsonian level against other major currencies, and the price of gold, which had at the end of 1974 approached $200 per ounce, was down to $125 per ounce, partly due to a general weakness of commodity prices and partly due to the fact that the IMF was thinking of selling off some of its gold.

Economic policy makers in 1974 were like pilots whose controls had been miswired. Normally when they saw inflation shoot up, they pulled back on the throttle to cool it down by reducing demand. But the inflation in 1974 had not come from too much demand; it came rather from too little supply. Tight money and fiscal measures to combat inflation by cutting demand did cut inflation below what it would have been, but in the face of negative supply pressures the cutback in demand threw output and employment into a sickening dive. By the time the restrictive demand policies were effectively reversed in 1975, the industrial world had slid into deep recession.

Banks Booming in the Midst of Chaos

In the midst of this maelstrom of confusing cross currents and inflationary recession, the commercial banks of the world were enjoying a boom in a new line of business—megalending to developing countries. They lent about $15 billion to non-oil developing countries during 1974 alone, and by the end of 1975 had loans out to developing countries totaling about $63 billion. That number rose by almost a third, to $81 billion, at the end of 1976. While many people thought the industrial world at mid-decade was falling apart, the banks obviously were convinced they had found a safe refuge from the storm that was swirling around them at home.

We have already talked about aspects of foreign lending that attracted bankers during the mid-1970s: the attractive economics of lending hundreds of millions abroad at a clip instead of only tens of millions at home; the relatively impressive performance by many developing economies after the oil shock, especially by Brazil, an oil importer; the tremendous quantity of funds to be recycled from OPEC to oil importers; and perhaps most compelling, the desire to invest in the "new economic order," which saw countries rich in natural resources—oil today, what next?—as the place to be.

In addition to these attractions abroad, forces at home were acting to push the banks out of domestic activity. An economy spiraling into recession as inflation accelerates does not create an attractive market for making huge loans —the massive placements needed to recycle oil billions. Demand for loans in the industrial world was weak, especially at the very high interest rates in the United States, and a lot of what there was arose from distress borrowing— businesses seeking funds to stay afloat, without much in the way of alternatives.

Loans to Mexico, Argentina, Brazil, Venezuela, and Korea or Taiwan— which were made at rates of up to two full percentage points above the London interbank rate—seemed like loans to winners instead of loans to tide over losers. Besides, the loans were in dollars and generally carried floating interest rates so the banks were protected from loss if inflation should spiral up further carrying interest rates with it. What was forgotten—not to be recalled until 1982—was that floating interest rates do not remove all risk from country lending. If inflation should fall more than interest rates, then the real burden of country debt is compounded by the double burden of higher real interest costs and lower commodity prices that mean a double whammy of less available foreign exchange with which to pay the higher real interest costs.

Economic Summitry Begins—Again

Amid the stagflation, exchange rate gyrations, and pervasive sense of gloom during 1975, the leaders of the six biggest industrial countries—Messrs. Ford, Miki, Schmidt, Giscard d'Estaing, Wilson, and Moro—decided—misery loving company as it does—to get together for the first of what was to be a continuing series of world economic summits. The feeling behind the summit was that nations must work together to pull the world out of a slump. The Economic Summit was heavily promoted and elaborately staged in an elegant setting thirty miles southwest of Paris at the Château de Rambouillet, where the world's most famous collection of tapestries is housed.

The outcome was anticlimactic—inevitably, since it was not very likely that after a decade of accommodating inflation, six men, however powerful, could get together over a pleasant weekend in the French countryside and undo what the oil sheiks had done. The usual communique was issued. The six countries that had led the industrial world into stagflation declared they would not accept another burst of inflation—perhaps having by then begun dimly to realize that it was supply starvation, not demand pressure, that had the economic machinery squeaking. Together they pledged—in anticipation of virtually every communique to follow subsequent summits—to avoid beggar-thy-neighbor protectionism and committed themselves to accelerating trade negotiations underway in the Tokyo round of the General Agreement on Tariffs and Trade (GATT), talks. And of course there was the obligatory pledge to join in urgent measures to help the Third World meet its deficits and to stabilize its export earnings—the banks were already there for the strongest of the Third World countries. Finally there was the usual Franco-American wrangling on exchange rates in advance of the meeting on international monetary reform scheduled for January in Jamaica.

The Rambouillet Summit was eerily reminiscent of a world economic conference held just over four decades earlier in London during June of 1933. That conference put forward proposals for coordinated world reflation, coordinated monetary and fiscal policies, a tariff truce and gradual removal of import and capital controls, and support for commodity prices. In the background, in 1933 as in 1975, especially from Britain and France, were pressures for the United States to lead the world out of recession with its own boom.

The eventual fruitlessness of the 1933 proposals frightened those who remembered them when comparisons were made with the immediate outcome of Rambouillet. In its commentary on Rambouillet, the London *Economist* recalled the fifty-year Kondratieff long cycle of slumps: the depressions of 1826 and 1829, the lengthy depression of 1873–96, and the Great Depression of 1929–35.[9] That theory—really more of a simple catalog of fairly regular economic slumps—together with Joseph Kitchin's short-range inventory cycle pointed to the 1975 slump and a possibly worse one four or five years ahead in 1979–81. The *Economist* left little to one's imagination:

There is greater reason to fear that the 1975 slump might prove to be another mishandled convulsion like that of 1929–35. The sequence could run from stockmarket crash (1929 and 1974) through financial crisis (1930–31 and 1974–??) to aborted recovery (1930 and 1975–76). If there is an abortion of the recovery over the coming 12 months, politicians may over-react and produce a great unstable boom in 1977.[10]

Of course, that is exactly what they, led by the new U.S. President Carter, did. What was missed by everyone—including the *Economist* in its historically accurate and sweeping discussion of Rambouillet and the economic events surrounding it—was that although the symptoms in 1974–75 resembled those in 1929–30, the disease was very much different. There was one clue, the behavior of prices: sagging in 1929–33, soaring in 1974–75 and beyond. The era that gave rise to Keynes's great *General Theory*, that discarded the glib presumption of Say's Law whereby supply created its own demand, spawned a generation of economists living and working across the fifty-year span between the low valleys of the Kondratieff cycle who kept looking on the demand side. They had learned that while supply does not necessarily create its own demand—at least not right away—insight in no way suggests the inverse, that demand creates its own supply. Looking back, the mistake is obvious. Looking ahead, it was not. The error was costly because it was the powerful demand stimulation of the 1976–78 period that resulted in validation of what otherwise would have been adjudged too much lending by banks to LDCs during the mid-1970s.

A Sigh of Relief: Easy Money

By the end of 1975 the industrial world was beginning to catch its breath after two years of dizzying speculation on consequences of the oil shock. It had been termed by five economists writing in *Foreign Affairs* at the end of 1974 "a financial problem of staggering proportions".[11] Amid fears that OPEC would buy up the rest of the world, which would bankrupt itself trying to pay for oil, oil shock had by the end of 1975, according to two Brookings economists, become "no significant threat to living standards or economic growth."[12] The senior partner in the Brookings study was Charles L. Schultze, who a year later would be tapped by Jimmy Carter to become chairman of his Council of Economic Advisors. Writing in the December 7, 1975, *New York Times* of the Schultze study under the headline "What Happened to All the Oil Billions?" Edwin L. Dale, Jr.—affectionately known as the unofficial Treasury mouthpiece—pronounced the oil shock "not the devastating blow to the world economy that many had feared" and, acknowledging that everyone was "a little worse off because of it," predicted "the problem will grow less as time goes on."[13] Apparently he believed the Rambouillet commitment to control inflation.

Rambouillet marked a turning point in one sense. Prior to that first economic summit gathering in November 1975, U.S. monetary policy had been growing very tight relative to that in Germany, Japan, and Britain. Afterward it began to ease sharply—perhaps with a nervous Republican eye toward what looked like disastrous congressional election results the following November and perhaps because stagflation had left those inclined to be inflation fighters feeling that they had got very little for their troubles. They had, in fact, been subjected to scorn—at home, by those who blamed them for inflation, and abroad, by those who said they did not like a strong dollar that meant inflationary depreciation of their own currencies against the dollars used by OPEC to measure their oil bills.

The unhappy experience with stagflation during 1974 and 1975 led many to look diligently—and in the wrong place—for explanations. Those at the IMF, in France, and in Europe in general who had never liked the idea of floating exchange rates conjured up the absurd notion of "vicious circles," whereby floating exchange rates were to drive an ever-widening wedge between two groups, one with self-perpetuating high inflation and the other with stable low inflation. A vicious circle of inflation, currency depreciation, and more inflation was to curse the hapless victims (Britain, France, and Italy), while the perpetrators (the United States, Japan, West Germany, and Switzerland), with comparatively low inflation rates, caught in a "virtuous circle," would experience ever-appreciating currencies. The vicious and virtuous circles idea appealed strongly to economists who thought like bankers—like Alexander Lamfalussy at the BIS, where the untidiness of floating had never been well received.

The vicious–virtuous dichotomy was formally put forward in the fall of 1976 by the Paris-based Organization for Economic Cooperation and Development and echoed in Washington at the IMF and in Basel at the BIS. It amounted to no more than a means of classifying countries according to which ones were at the time high inflation (vicious) and low inflation (virtuous). By the summer of 1977 the United States would have to be reclassified from "virtuous" to "vicious" since by August the dollar had slid sharply against the German mark, the Swiss franc, and the Japanese yen.

The vicious-circle story was put forward largely by Europeans as part of an effort to force reflationary policies on the Americans. In a way it was a clever bargaining ploy because even if the pressure for reflation worked, it would mean some sharp exchange rate changes as dollar depreciation reflected U.S. reflation and the vicious-circle advocates who really disliked floating exchange rates would have more "currency gyrations" to complain about.

Demand Does Not Create Its Own Supply

The Pace Quickens: "Locomotive"

The Europeans and the Japanese who had in 1975 hoped for more expansionary policies from the United States got more than they had bargained for from Jimmy Carter and his economic team. The Carter people came to Washington in January of 1977 all revved up to get things moving with activist policies. In a December 27, 1976 *Newsweek* interview Michael Blumenthal, Carter's designated Treasury Secretary, did "not believe that at the present time, with our economy operating below capacity [that] the risk of inflation is very great as long as we remain cautious in the choice of the kind of stimulative policies we opt for. . . ."[14] Blumenthal still had not seen that a shortage of demand was not the problem.

The conservative Ford team—William Simon and Ed Yeo at Treasury, Arthur Burns at the Federal Reserve, and Alan Greenspan at the Council of Economic Advisors—was replaced (Burns, a year later when his term expired) by a group of activists ready to pump up demand and pull the world out of its economic doldrums. They believed that the oil shock was behind them. The "locomotive theory" whereby the U.S., German, and Japanese economies would pull the world to recovery was propounded with the support of facile simulations conducted by Lawrence Klein—later a Nobel laureate—and his associates at the Wharton School. Their model showed "clearly" that all the United States had to do was expand along with Germany and Japan and the world economy would prosper. Right after Carter's inauguration Vice President Walter Mondale was sent on a world tour together with the Undersecretary of State for Foreign Affairs, Richard Cooper—a Yale economist in Carter's brain trust—and Treasury Assistant Secretary for International Economic Affairs C. Fred Bergsten—a perennial Washington fixture who had come along with Charles Schultze from the Brookings Institution, which was reputedly a storehouse for democratic policy types. Bergsten and Cooper pressed strongly their locomotive idea, urging bigger budget deficits and easy money on the Japanese and Germans who—having by then understood that supply, not demand, was the problem—were having none of it. The U.S. locomotive roared out of the station on its own.

The U.S. money growth, which had risen from 4.9 percent in 1975 to 6.7 percent in 1976, jumped to 8.1 percent in 1977. The federal budget deficit, which had been scheduled to come down sharply after the drain of Vietnam had ended, went from $45.2 billion in 1975 to $66.4 billion in 1976 and stayed at $45 billion during 1977, swollen another $9 billion to $54 billion by off-budget outlays.[15] By summer the dollar was back in the doldrums amid charges that the cigar-chomping Treasury Secretary Blumenthal was "talking down" the dollar with an "open mouth" policy.

In truth, the U.S. locomotive was doomed before it left the station. During the period prior to 1976 when U.S. monetary and fiscal policy had been restrictive, the other U.S. policy of keeping oil prices low to "cushion" the impact on the U.S. economy had not resulted in enough imported oil to push the U.S. balance of payments into deficit. But when Carter's engineers slammed open the throttle on the locomotive in 1977, it began to suck in huge quantities of $10-a-barrel oil. It seemed cheap, and, by worldwide standards, it still was. The trade deficit accelerated to $2.8 billion per month by June, reaching a 1977 total of over $30 billion and a 1978 total of $33 billion. Oil prices shot up by over 9 percent in 1977 and, even more impressively, prices of non-oil primary commodities soared by 20 percent in that year (after having collapsed at an 18 percent rate during 1975).[16]

Lending Shifts

Commercial bank lending to developing countries grew less rapidly in 1977 than it had the year before, and U.S. banks' participation in the growth fell from 48 percent in 1976 to 28 percent in 1977. The U.S. banks had plenty of lending opportunities at home, where the Carter administration's stimulative policies caused growth amid a need to buy a lot more oil. The big banks in Germany and Japan saw that oil and other primary product exporters would benefit from the U.S. locomotive and moved to invest heavily abroad, especially in Latin America, while typically cutting back their commitments in the United States. The trend continued in 1978, when bank lending to developing economies soared by almost $40 billion, to $131 billion, with U.S. banks accounting for only 14 percent of the total growth. U.S. growth rates of over 5 percent from 1976 through 1978, accompanied by the tremendous boom in mortgage lending, presented more opportunities at home, although financing oil imports was absorbing more and more dollars.[17]

The 1977–78 period was a significant one for U.S. banks. At that time they were evaluating returns from their first burst of LDC loans during 1974–76, when they had recycled the first huge surge of petrodollars. The results were encouraging. Thanks to the strength of commodity prices, the borrowing countries were able to meet scheduled debt-service payments. The unexpected surge of inflation pushed down rates of return on fixed interest loans during the locomotive years—many lenders had mistakenly believed the Rambouillet pledges of no inflationary policies so that by 1977 inflation-adjusted returns on many fixed interest investments were negative. But the loans to LDCs carried floating interest rates, tied to inflation-sensitive market rates that floated up with inflation, thereby preserving the attractive real returns on bank loans to LDCs. As a result shares of bank earnings from LDC loans rose and by 1979 the banks were ready to undertake a sharp reexpansion of lending to developing countries. During 1979 bank lending to developing countries soared by $40 billion, with the share of U.S. banks rising back up to 25 percent.

Lessons of the First Oil Crisis

The mid-1970s were a disturbing and disorienting period especially for U.S. policy makers and economists. The Europeans and Japanese, more used to shocks from the outside and the scarcity of critical raw materials, had by 1976 seen the handwriting on the wall. Oil conservation must be accelerated, and the way to do it was to put oil prices up very high—to move oil consumers from the "one-bedroom" to the "studio" stage. U.S. politicians just were not ready to shove Americans out of their four-bedroom houses and huge cars that fast. For an Italian or a Frenchman owning a Fiat or Renault that gets fifty miles per gallon, largely because gasoline had cost over $2.00 per gallon, $2.50 a gallon is a bearable shock. For an American with a ten-mile-per gallon Buick and 30-cents-a-gallon gas, $2.50 a gallon would lead to fighting in the streets. U.S. politicians knew this and held back. The policy was viable until the U.S. economy began to expand after 1976. Then it sucked in oil and gushed out dollars instead of goods to pay for it, at a rate that made another explosion of oil prices inevitable.

The immediate symptom of the dollar gusher to satisfy the greedy U.S. appetite for oil was a steadily weakening dollar. The dollar's tailspin kept most eyes glued on the foreign exchange markets and the U.S. balance of payments as the crisis area. That was a mistake. The real crisis lay with U.S. dithering on energy and continued subsidies to oil consumption.

The message that observers learned from the first oil shock was that U. S. leadership would have been required to resist the inflationary pressures it set in motion and that such leadership was lacking. Resistance to inflationary pressure during 1974 and 1975 had crumbled in the face of charges that the

United States was precipitating a worldwide recession. By 1977 the turnaround was so sharp—far sharper than many in Germany and Japan wanted—that it seemed obvious that another oil price surge would not produce even a pretext of U.S. resistance to its inflationary consequences. Accommodation would be prompt—in fact, so prompt that the big fear was a serious acceleration of worldwide inflation. Europeans and Japanese, fearful of this outlook, put so much pressure on the dollar that by November 1, 1978, President Carter was forced to reverse the expansionary policies set in motion just twenty-two months earlier.

Lessons Ignored

The Carter administration was slow to acknowledge that its expansionary policies of demand stimulation and subsidized oil prices were totally inappropriate, representing as they did a dangerous, inflationary inversion of Say's Law. The dollar's weakness—very sharp by December of 1977—reflected comprehension of the dangers by those outside of the United States. The White House and Treasury responded with what amounted to a charge that the markets were "crazy"—driven by "irresponsible speculators"—and that a turnaround would come soon.

To the Europeans and Japanese, the persistence of Carter's expansionary policies well into the fall of 1978 was nothing short of incredible. So frightened of a collapse of the dollar standard were the usually antagonistic Germans and French that Messrs. Schmidt and Giscard d'Estaing buried the hatchet to put together in the spring of 1978 the European Monetary System. That system was nothing other than a hedge against collapse of the dollar as a viable monetary standard.

By October the Carter White House finally began to think about some anti-inflation medicine. They were flying blind. Burns had been replaced at the Federal Reserve by G. William Miller, a man with virtually no experience with the delicate problems of managing monetary policy. Blumenthal and Tony Solomon, the Undersecretary at Treasury in charge of international matters, had used up what credibility they had in pursuing two years of policies that obviously would ruin the dollar while denying anything was wrong. It was the same old story. International finance was of no interest to the engineer who was President, and his Treasury Secretary—no expert himself—was not about to bother him with it, especially in view of his poor personal relationship with Carter and "the White House kids."

The almost incredible degree of incompetence with which the Carter people handled the economy in 1977 and 1978—persisting with demand stimulation in the face of a supply problem, as if by beating harder a horse that had collapsed in harness they could bring it back to life—had not been—could not have been—anticipated. The policies that would never have been followed by a responsible, informed U.S. leader created remarkably favorable conditions for borrowers, who began a rush to borrow more. At home those who bought real estate, gold, rugs, or anything tangible realized tremendous rewards while buyers of U.S. government securities were left to feel like fools, reaping outright losses with interest rates well below inflation. Losses on fixed rate mortgages began to threaten the solvency of savings and loan institutions. In contrast, the experience of banks with their loans to LDCs was very favorable because of their floating interest rates and because the prices of goods sold by the borrowers went up so fast that they had no trouble meeting debt-service payments on existing loans while contracting for new ones.

Another Expansion Goes Too Far

The crisis in October 1978 was more devastating than the May 1971 run out of the dollar, partly because the United States was out of options save to reverse its inflationary policies but more fundamentally because the men in charge were in way over their heads, and everyone—especially the Europeans —knew it. In 1971 it had been Connally and Volcker at Treasury, Burns at the Fed, and Nixon—for all of his faults an experienced, canny leader—in the White House. In 1978 it was Blumenthal and Solomon at Treasury, Miller at the Fed, and Carter in the White House. The Europeans had seen these men up close over the past two years, had taken their measure and decided— correctly, as it turned out—that they were totally incompetent to set a viable course for U.S. economic policy. The collapse of the dollar in October 1978 signaled a collapse of confidence in U.S. hegemony on world economic matters.

In a televised speech on October 24, Carter's first response was to talk about future budget deficits and wage and price guidelines. No mention was made of monetary policy or international economic policy. The dollar collapsed along with the stock market.

On November 1, 1978, a chastened President Carter personally announced a 1 percentage point increase in the discount rate—the largest single increase in almost five decades and the first ever announced by a president. (Such announcements are usually left to the chairman of the Federal Reserve Board.) In addition, the United States agreed to absorb the tremendous quantity of dollars that the German, Japanese, and Swiss governments did not want to hold by issuing Carter Bonds, denominated not in dollars but in German marks, Japanese yen, and Swiss francs. This humiliating concession acknowledged that until further notice U.S. IOUs were unwelcome as far as responsible foreign governments were concerned.

The intended effect of the locomotive—so badly sidetracked by inflation—had been to get Europe and Japan moving faster. Fifty years earlier Sir Montagu Norman, governor of the Bank of England, and Hjalmar Schacht, president of the German Reichsbank, had quietly sailed on the *Mauretania,* arriving in New York on Friday July 1, 1927, to meet with Benjamin Strong, head of the Federal Reserve Bank of New York. Their aim—like the European aim at Rambouillet—had been to urge U.S. expansion to rekindle recovery in Europe. As it did in 1977 and 1978, so in 1927 and 1928 did the expansion go way too far, fueling a speculative boom that became a crash. In 1929 the boom was in the stock market until October 1929. In 1977 and 1978 the boom was in commodities, houses, and bank investments in developing countries that supplied commodities. The twenties boom really began to unravel in 1931, with the failure of Kredit Anstalt, which pushed Britain off the gold standard. The 1977–78 boom really did not begin to unravel until the U.S. Federal Reserve, at first abortively in the fall of 1979 and finally for real in the spring of 1981—after false starts in 1969, 1971, 1973, 1975, 1978, 1979, and 1980—made the move to control inflation that almost no one believed it really would make. United States adherence to a policy of inflation control—in deed and not just in word—from March 1981 through August of 1982 was the biggest shock to the world economy since the first oil shock in 1973. Unlike the negative supply-side shock from oil, it was a negative *demand* shock, the last thing for which banks and businesses were prepared. The floating interest rates on loans to LDCs were not protection against a leap in real interest costs and a collapse in commodity prices. The loan contracts called for attractive terms under any eventuality for the banks, but they were not worth much if the borrower turned out to be totally unable to meet those terms. By August of 1982 the painful truth of that fact had brought Mr. Silva-Herzog to Washington.

Chapter 7

Imprudent Moments

We lost the Eurodollar market to the United Kingdom when we passed that dumb Interest Equalization Tax. We lost the foreign exchange market to
the U.K. and the bond market. I'd like to get the
market back. . . .
WALTER WRISTON, in a July 1978 interview

The world is a Catch-22 situation.
WALTER WRISTON
in a September 1983 interview

THE gold bubble of October 1960 proved to be as much of a watershed for
commercial bankers as it was for their cousins in the finance ministries and
central banks whom it set feverishly to work on holding together the Bretton
Woods system. As a direct result of jitters in the gold market and the currency
turmoil of March 1961, when revaluations of the German mark and the Dutch
guilder effected the first disguised postwar dollar devaluation, John F.
Kennedy's new Treasury Undersecretary for Monetary Affairs, Robert V.
Roosa, began with his boss, Treasury Secretary Douglas Dillon, to devise ways
to stem the net outflow of dollars. More dollars flowing out of this country that
ended up either at foreign central banks or in private coffers that could empty
rapidly into foreign central banks all represented volatile claims on the dwindling stock of gold in the United States. Such dollar flows were tabulated every
quarter by the Commerce Department as the "liquidity balance" in the U.S.
balance of payments accounts. The liquidity balance showed that although
U.S. commodity exports comfortably exceeded its imports—by an average of
$5 billion annually during 1960–64—beginning in 1958 capital outflows, U.S.
investment abroad, began to more than swamp the trade balance surplus.
During 1960, even with exports exceeding imports by $4.9 billion, net capital
outflows were so large that the liquidity deficit recorded an increase of $3.7
billion in foreign hands that year. That brought to $21 billion the stock of

dollars held outside the United States that could be converted into gold whenever confidence in the dollar started to wane. In 1960 the U.S. gold stock stood at $17.8 billion. The 1961 liquidity deficit shrank to $2.25 billion, but it rose again in 1962 to $2.86 billion. Meanwhile the U.S. gold stock dwindled by almost a billion dollars in 1961, to $16.9, and by another $900 million in 1962, to $16.0 billion, over $8 billion less than the $24.3 billion in dollar assets held outside the United States.[1]

The first efforts by Kennedy, Dillon, and Roosa to deal with the unaccustomed problem of "defending the dollar" were constructive. A drive to raise exports included more resources for the Export-Import Bank, which helps U.S. exporters offer attractive borrowing terms to their foreign customers in need of trade financing. The United States also pressed for multilateral tariff reductions through the General Agreement on Tariffs and Trade. In 1962 the Federal Reserve turned its attention to the capital flows problem with Operation Twist. The idea was to twist the yield curve describing interest rates on debts of differing maturities from short to long, so that short-term rates would stay high and attract capital inflows while long-term rates would stay low to encourage investment. It did not do much good because lenders quickly moved loans from the long to the short end of the market thereby erasing most of its impact. By the end of 1962 Roosa and Dillon were running out of "nice" ways to deal with the hemorrhaging of dollars and the patient was getting low on reserves. What was really needed was a whole new system that let the dollar price of gold rise with some exchange rates adjusted, but in 1962 such radical surgery seemed far too risky. The doctors opted instead for a less radical move to stop the bleeding. One of its major unforeseen side effects was to move U.S. banks, for the first time, into the business of direct term lending abroad, the forerunner of the direct bank-lending explosion that followed the first oil crisis.

The Start of Direct Bank Lending Abroad

On July 18, 1963, President John F. Kennedy reluctantly imposed a "temporary" Interest Equalization Tax (IET) of 1 percent on all foreign securities sold in the United States. IET meant that if an American investor lent abroad by purchasing foreign securities, the after-tax return would be about 1 percent lower than on alternative uses of funds. With interest rates in the 3 to 5 percent range, as they were in 1963, IET amounted to a prohibitive tax.

As with all stop-gap measures, IET had its loopholes. First, it exempted Canada and Japan—the former for reasons of "good neighborliness," the latter as a reward for willingness to accumulate and hold, without grousing about possible conversions into gold, more and more dollars. Besides, the Japanese liked to borrow in dollars since dollar payables provided a nice hedge for the dollars they received for exports to the United States. Unlike the Germans and some others who required U.S. importers to obtain foreign currency to pay for their shipments of Volkswagens, Mercedes, and Volvos, the Japanese obligingly invoiced their U.S. customers in dollars, thereby relieving them of the problem of dealing with unfamiliar foreign exchange transactions. This practice was continued by the Japanese until the 1970s, when heavy losses on dollar receivables and diversification through the eurocurrency markets to include borrowing in other currencies made it less attractive.

The biggest loophole in IET was a gift to the U.S. commercial banks that must have set them to thinking that maybe a democratic president was not so bad after all. While IET effectively cut foreigners off from U.S. bond and equity markets, it exempted from the tax commercial bank loans of less than three years' duration. Borrowers in Europe and Asia, together with those in the more advanced developing countries that had been relying on capital flows from the United States, were thrown into the arms of U.S. banks, banks that began a profitable business in thirty-five-month term loans. Previously U.S. banks had largely confined themselves to very short term trade financing activities. The banks' new term lending virtually replaced the outflows that IET was meant to prevent. The 1963 liquidity deficit did drop to $2.5 billion from $2.9 billion in 1962 but rebounded to $2.8 billion in 1964.[2]

Eurodollars Stimulated by U.S. Capital Controls

IET was to have expired at the end of 1965. Instead it was extended by a Vietnam-burdened Lyndon Johnson to the end of 1969 (and thereafter to the end of 1973, finally expiring in January 1974). In addition, the 1965 version of IET covered bank loans of more than one year's maturity, thereby ending the popular thirty-five-month term loan. In addition to extending IET, an ever-more-nervous U.S. government—for in 1965 dollar holdings by foreigners had reached $30 billion, over twice the $14 billion stock of U.S. gold—imposed the "Voluntary" Foreign Credit Restraint Program (VFCR) administered by the Federal Reserve. The Fed with its regulatory responsibilities has many ways to cause banks to "volunteer" cooperation and did so under VFCR, which limited acquisition of foreign assets by U.S. banks and financial institutions above established base levels. A mandatory set of Commerce Department controls on foreign direct investment was added in 1968, completing a panopoly of ill-advised restrictions on U.S. investment in a growing world economy.

The capital controls program, as the web on restrictions of U.S. foreign investment came to be called, was largely ineffective. The banks simply expanded their London branches and subsidiaries and participated in the immense growth of international lending through the eurodollar market. Other means to subvert the controls were found, as suggested by the sharp rise in 1969 of the "errors and omissions" line in the U.S. balance of payments accounts, which measures unrecorded international transactions. Previously it had run about $500 million; by 1969 it was $2.6 billion.[3] The U.S. banks demonstrated considerable ingenuity in assuring their participation in the growing eurodollar international lending market during the 1960s. From the host cities' viewpoints the primary impact of the capital controls programs was to move a great deal of revenue-generating banking activity from New York to London. The international capital markets grew up to achieve a supranational character that the official sector of the international financial system had groped for tentatively in 1944 and then failed to achieve in the crucial two decades that followed. The incredible Eurocurrency System reached $200 billion by 1971, operating largely out of London but also out of all the world's major financial centers. Had the British tried to quash it—as they wisely did not do—it would merely have resurfaced somewhere else where friendly local authorities, grateful for the ancillary business and revenues generated, would welcome it.

U.S. Banks' International Departments Expand

The U.S. program of capital controls—designed to quash capital flows out of the United States—did just the opposite. It introduced U.S. banks and bankers to international business just when the transportation revolution brought about by the introduction of commercial jet aircraft and worldwide electronic communications made it possible to think about borrowing and lending anyplace in the world. The international departments at U.S. banks, once backwaters, began to grow after IET in 1963 and, after 1965, to spill over into operations in London, Tokyo, Singapore, and Frankfurt. Like young athletes who begin training for the Olympics eight years before the games, their muscles were honed and ready for the great challenge when in January of 1974 the capital controls programs were ended just as a quadrupling of the price of oil had produced almost overnight the biggest increase in the needs of trans-national borrowers and lenders in the history of the world.

For U.S. bankers 1974 was a year of rapid transition. At home they faced stagflation, with inflation roaring along at 12.2 percent and negative growth of 1.4 percent. Both of these indices of economic discomfort were by far the worst of the postwar period. Meanwhile some $68 billion in oil surpluses had to be recycled to developed as well as developing countries; the latter were still managing an average growth rate of 5.5 percent against the former's average of a minuscule 0.6 percent. Little wonder that the "big six," Bank of America, Chase, Citibank, Morgan, Chemical, and Manufacturers Hanover, all banks where the oil billionaires liked to deposit their funds, began a rapid expansion of lending to developing countries. The U.S. government provided its bankers with plenty of encouragement, lecturing the banks about their recycling "re-sponsibilities" and—although it came as a coincidence, it had been planned well in advance of the oil shock—conveniently terminating controls on capital outflows on January 29, 1974.

Even before 1974, banks—including the big ones in the United States work-ing through London branches—had begun lending more to the booming devel-oping countries like Brazil, Taiwan, Mexico, and South Korea. By 1973 the IMF and World Bank reported bank lending to developing countries at about $34.6 billion. That total soared 40 percent during 1974, to $48 billion, and by another 30 percent, to $62.7 billion, by the end of 1975.[4] The U.S. banks that had gotten their head start on direct term lending to LDCs a decade earlier, thanks to inducements of IET loopholes, had the jump on their foreign compe-tition at first. In 1975 they held $34.3 billion of the LDC debt, 54.5 percent of the total and the highest share they were to reach. The year 1976 was another boom year for the banks, with total lending to LDCs rising another 29 percent and U.S. banks taking 53 percent of that increase to bring their total LDC lending to $43.1 billion.

The significance to U.S. banks of the mid-seventies LDC lending boom can be seen from the sharp increase in their share of earnings coming from interna-tional business. The share of earnings attributable to international business was: 40 percent in 1976 versus 21 percent in 1972 for Bank of America; 72 percent versus 54 percent for Citibank; 78 percent versus 34 percent for Chase Manhattan; 56 percent versus 29 percent for Manufacturers Hanover; 53 percent versus 35 percent for Morgan Guaranty; and 44 percent versus 14 percent for Chemical.[5]

Citibank and its aggressive chairman, Walter Wriston, were at the head of the pack, leading the charge toward more lending to LDCs. By 1976 Citibank was getting 13 percent of its total earnings from Brazil, compared to 28 percent from the U.S.[6] The tremendous increases in lending to LDCs reflected the remarkable profitability of such loans compared to domestic lending. Much of the domestic lending had been done at fixed interest rates of 6 or 7 percent; the banks were stuck with these lower rates over the life of such loans, which in some cases was more than ten years. In contrast, loans to LDCs were made at premiums of 1.5 to 2.0 percentage points over interbank rates in London and had floating rates so that bank earnings from LDC loans would be protected if interest rates rose even more later on; no one seemed worried about what would happen if rates went so high that the LDC's *could not* pay. While in 1977 the loans to LDCs accounted for only about 8 percent of the total assets of the nine largest U.S. banks, the banks were on average deriving at least half of their earnings from these loans. By being the most aggressive banker in the field of LDC lending in the 1970s, Walter Wriston led Citibank to a position as New York's largest and most profitable bank, handily beating out rivals at Chase and Morgan.

Banks' Exposure to LDCs Grows

The aggressive move by the big banks to expand profitable LDC lending of course had its price. By the end of 1977 the nine largest U.S. banks had 163 percent of their net worth invested in non-oil LDCs, with about half of that in Argentina, Brazil, and Mexico alone. Asked in a 1978 interview with William Clarke why he didn't feel vulnerable, Wriston—sounding very much like a salesman ducking a tough question with some jargon—replied: "We run what we call our actuarial base, which means that we don't know where trouble will come from so we never have too much exposure anywhere, but we have some chips in every game in town."[7]

Citibank had 100 percent of its net assets committed in Brazil alone, suggesting that the diversification principle espoused by Wriston in the summer of 1978 had somehow fallen by the wayside; after half a decade, pushed aside by the heavy pressure to keep making more profitable loans to LDCs than anyone else dared in order to stay ahead of the competition. Wriston, Citibank, and its competitors were pushed onward by a belief that growth of world trade would continue unabated. Said Wriston: "It's been popular to say world trade will slow down. I don't believe it. It's doubled in the last 10 years [1968–77]. I think it will double in the next ten."[8] Half a decade after the end of 1977, world trade had grown by only 12.2 percent, having actually shrunk by 2.5 percent during 1982. Little wonder that by 1983 Citicorp—Citibank's parent corporation—found itself with a 24 percent rise in net income after the first nine months of the year and a 21 percent drop in the price of its stock between May and November of 1983.

Investors in Citicorp's stock either had done their homework or were listening in droves to investment counselors who had. Citicorp's 24 percent increase in earnings reflected improved consumer banking results without reflecting much in the way of actual and possible effects of arrearages of payments on LDC loans. Loans that are ninety days in arrears are declared nonperforming, which means that six months' interest on such loans must be deducted from earnings. Citicorp had about $5.6 billion in loans outstanding to Brazil at the end of 1982.[9] At average interest rates of 13 percent, such loans would contribute $728 million annually to Citicorp's earnings. Having to deduct six months' earnings on Brazilian loans alone would cut $364 million from its earnings in 1983—49 percent of its total 1982 earnings and 41 percent of its estimated $878 billion of earnings in 1983. Little wonder its stock was dropping even with higher earnings, pushed down by the mid-1983 jitters over massive rescheduling of even more massive Brazilian debts.

Citicorp was not alone in facing heavy write-downs on Brazilian loans, where arrearages had mounted by the fall of 1983 to over $4 billion. Among Citicorp's fellow New York banks, Manufacturers Hanover had $2.0 billion in Brazilian loans, and development-loan-oriented Chase, $2.6 billion—both exposures comparable to Citibank's in view of their smaller net worth. It is likely that some judicious rolling over of loans had been required to avoid triggering the "nonperforming" alarm bell on Brazilian loans. We have already seen that loans that had a sixty-day nonpermanence clause had in September 1983 been relaxed to a ninety-day clause to avoid the costly nonperformance designation. With respect to their LDC clients, the mighty banks were in the position of a bomb squad disarming a time bomb. Top priority at the moment was to disarm the nonperforming fuse—or worse yet, the default fuse—before it ignited the debt bomb. At such a critical moment capturing the bomber—like reflection on fundamental causes of, and long-run solutions to, the massive overhang of developing country debt—was a secondary consideration.

Concern was focused on Brazil during the summer and autumn of 1983, for it was the largest single debtor, and a refinancing package was painstakingly being assembled. But the flagship of the new economic order was by no means the only country that worried the banks and, even more intensely, the Federal Reserve System and the Comptroller of the Currency jointly responsible for certifying the financial health of U.S. banks.

By May of 1983, after the exposure of U.S. banks had grown further with the multiplication of loan reschedulings, more detailed figures emerged. The "big six" had lent $37 billion to the governments of "the big five," (Mexico, Brazil, Venezuela, Argentina, and Chile) alone. Additional loans totaling another $80 billion were outstanding from these and other U.S. banks to other developing countries. The $37 billion that the six largest U.S. banks had lent to five volatile Latin American states represented an average of 190 percent of their net assets. A bank that loses 100 percent of its net assets is bankrupt, or "insolvent," to use the more sterile and less descriptive term preferred by financial specialists. These numbers meant that if 52 percent of the loans to these five countries were to go bad, the six largest banks in the United States would be bankrupt and, barring some major rescue effort, would cease to exist. The U.S. financial system and very likely that of the world would be unable to withstand such a shock. Nor were these anxieties confined to the United States. The Bank of Tokyo, Japan's largest international bank, had 80 percent of its net assets at risk in Mexico alone. Exposures of major West German banks to Poland's $26 billion in debt were even higher.

Some Principles of Banking

The basic economics of banking is such that a bank that loses even, say, 30 percent of its net assets is severely crippled. A large bank—one among the world's five or six largest—will have about $100 billion in total assets. Its net assets—what is left over after subtracting liabilities to depositors—will be about $4 billion. Its annual earnings typically run 10 to 15 percent of net assets, or between $400 million and $600 million. A write-off of something like 30 percent of net assets, or $1.2 billion, means a loss equal to all of the bank's earnings for two or three years. An equivalent loss for a middle-income U.S. family with an annual income of $35,000 would be total destruction of their $100,000 home with no compensation from insurance or any other source and with their having pledged it as collateral in their business.

Examine the exposure of the six largest U.S. banks in Brazil alone, and numbers of more than double the household-home-loss magnitude confront us. By the end of 1982, they had lent Brazil $14.1 billion, a figure representing more than 72 percent of their combined net assets. By the late fall of 1983, Brazil was more than $4.0 billion behind in payments on its loans, heading for even larger arrears by year end unless loans could be rescheduled on a massive scale. Boards of directors at the "big six" must have been beginning to wonder how much longer they could keep pouring huge quantities of additional loans required in the rescheduling packages—Mexico had taken $5.0 billion; Brazil wanted $6.5 billion; Argentina and the Philippines wanted repayment moritoria, additional money, and easier terms that all the borrowers would demand if they were granted to one.

The big money-center banks in New York and London that were facing these worrisome problems in 1983 and 1984 seem remote to most people who live in smaller cities or towns and deal with much less imposing banks. People's banks, women's banks, farmers' and mechanics' banks, and merchants' banks all present themselves as special institutions interested in serving individuals and small businesses. In a way they are the retail end of the banking business, while the big banks are the wholesalers. But heavy losses at the wholesale end of any business mean problems at the retail level as well. The smaller hometown banks, usually called "regionals" to distinguish them from the big "money center" banks, get most of their funds for local lending from the money centers. Money that the big banks must put aside for loan loss reserves or for actual loan losses is unavailable for investment or lending to the regionals. Because it no longer serves as an earning asset, it represents an increase in the costs for the big banks that is passed on to regionals in the form of higher borrowing costs on the funds they get from the money centers, costs that are in turn passed on to the regional banks' customers.

Factors such as unanticipated losses on LDC loans that cut into the profitability of money center banks affect everybody's credit cost. The loan losses mean that the banks will turn out to have underestimated their costs, and the reassessment can produce some large shocks. A bank that takes in a deposit of $1,000 and pays 10 percent interest to the depositor is required by law to hold an average of about $100 out of the $1,000 in nonearning reserves—the experience of the last two centuries is embodied in this enforced degree of caution. That means the bank has to earn 11.1 percent on the remaining $900 in order to earn the $100 in interest it has to pay to the depositors from whom it has borrowed the money in the first place. If it adds a markup of 1.1 percent for its own profit, the lending rate ends up at 12.2 percent. If the same bank is facing heavy loan losses and has to put aside an additional 5 percent as loan loss reserves, then with a total of 15 percent of asset as nonearning reserves it has to earn the $100 on $850 to pay depositors their 10 percent. That takes an 11.76 percent rate of return, which, after a markup of 10 percent, becomes 12.94 percent. Having to put aside 5 percent against loan losses adds three-quarters of 1 percent to the rate the bank charges its best customers. That can easily translate into a full percentage point on such retail business as mortgage loans. A rise in the mortgage rate from 12 to 13 percent adds about $1,000 a year to the interest charges on a $100,000 mortgage. In other words, it is not very hard to see how events in Brazil can have a big affect on an American family planning to buy a house in Seattle or a small lumberyard in Chapel Hill; nor is it hard to see why interest rates remained surprisingly high in 1983 as banks struggled to build reserves. Experience with high rates in the early 1980s has shown that mortgage demand dries up rapidly at rates over 12 to 13 percent simply because most families cannot make the monthly payments, and so by the end of 1983 the housing industry was showing signs of slowing down earlier than usual in an expanding economy.

The world's biggest banks may not have been worried that families in Omaha postponed buying a house because they had to increase their emergency reserves. But they had to worry about the fact that setting aside more of such reserves means that the banks' managers are being forced the lower earnings potential to deal with an unforeseen increase in risk attached to LDC loans.

Too Little Information?

The banks that had reached this state of affairs by 1982 may well have been surprised by the figures on the total exposure of all banks. Half a century before, as the financial system tottered in 1931, no less than Montagu Norman, governor of the Bank of England and pope of the prewar world economy, had observed with enviable detachment: "had the various lenders all been aware [of the foreign loans coming out of London], such loans would have been quite out of question."[10] Revelations about that year's sharp rise in total short-term lending to developing countries may have been an especially rude shock, but the banks must have known all along the exact extent of their own exposures. The boards of directors at Manufacturers Hanover, Chase Manhattan, and Citicorp could all look at the balance sheet and see that by 1982 their Latin American exposure alone had passed twice their net worth. Commitments like this had not been undertaken by naive or somnolent bankers but were sought anxiously as profitable loans to the most secure class of debtor, governments themselves. By 1982 almost half of the $3.3 billion in profits realized by the ten largest banks in the United States came from their international business; ten years earlier it had been only a tiny fraction. The aggressive new breed of international bankers saw the very cream of this business as huge loans to the governments of rapidly developing nations—the vanguard of a new economic order.

Imprudent Moments

A Banking Revolution

The new breed of bankers personified by Citibank's Walter Wriston recognized the basic truths about lending to governments of the developing countries. Those governments acted as intermediaries between huge consortia of diverse domestic borrowers and the big banks of the developed world. But as we have seen with a large number of separate projects under a government umbrella, the bankers saved themselves the cost of carefully investigating each project to determine its profitability and potential to repay borrowing, assuming they even had the language skills and technical and regional experience required to do so. Rather, the bankers could just deal with the governments guaranteeing these debts, pledging as collateral their ability to tax their citizens. The pledged word of a great republic like Brazil is very imposing at the time it is given, and somehow it induced a retroactive belief that governments do not default on their debts.

In addition to seeming formidably safe, the big loans to developing countries were, as we have seen, highly profitable, largely through the economies of scale inherent in bank lending operations. An idea of just how profitable they were —over and above the attractive high interest rates they earned—emerges from a few simple calculations. Fees for arranging the loans averaged about 1 percent of their value. The difference between the bank's cost of funds and the lending rate are expressed as "spreads," percentages of the value of the loan —which, while typically lower for large borrowers, vary a good deal less than the size of the loan. One percent of a $200 million loan is $2 million, while 2 percent of a $100,000 loan is only $2,000. It does not require anywhere near one thousand times the effort expended on a $100,000 loan to arrange a $200 million loan, and even if syndication costs should be large, they can be covered by additional fees assessed "up front" (before the loan is granted).

People with checking or savings accounts at the big money center banks like Chase or Citibank are constantly tormented by the implications of these compelling scale economies in banking. The huge banks do not want small depositors or their loan business, but they are required by law to provide such services to individuals—services that contribute little, if anything, to profits compared to the big lending operations. The loyal Citibank customer who sees himself as a substantial depositor with $25,000 to $50,000 in various accounts is just an annoyance to the bank. When he or she seeks a personal loan of, say, $30,000, the rate quoted is often 50 percent above that quoted by a smaller, retail-oriented bank. In effect, the big bank is saying that it could use the resources it takes to lend $30,000 to a thousand average individuals, with all the actuarial probability that some of the thousand will run into difficulties and give the bank costly trouble, to lend $30 million to a safe big borrower. And 1 percent of $30 million is $300,000.*

There is no doubt that massive loans to governments of developing countries are profitable to make. But are they safe? More concretely, will they be repaid? Will the interest come every quarter it is due?

These questions arise from the revolution in banking that led to the boom in direct bank lending to governments of LDCs, itself the product of a transformation in the processing of data on both sides of the loan market. Prior to the 1970s lending to LDCs was dominated by investment bankers—that branch of the profession which instead of acquiring liabilities of the borrower arranges for other people to acquire them. Investment bankers generally worked on a specific undertaking—a nineteenth-century railroad or a twentieth-century hydroelectric project—investigating in detail, determining soundness or lack thereof, and, if convinced, acting as intermediary to place the bonds that funded the project. Their profit came from a small markup for selling bonds to investors. Investment bankers had an incentive to do their homework carefully on projects they elected to underwrite since in practice they were the initial buyers of the securities to be marked up for prompt resale. If word leaked of a dam's weakness or if past railways had not paid their bondholders off, investment bankers stood to lose all of their profits by having to sell the bonds at a loss. Sometimes investment bankers got too close to their clients and acquired too much of their paper for their own account, thinking they were seeing opportunities that others were missing. If this proved wishful thinking, then ruin might follow: in 1890 the great house of Baring Brothers, London, crashed when its Argentine investments went sour.

But the investment bankers, professionally equipped to sell company-sized issues of securities to the rather specialized population of institutions and rich people that forms their constituency, were not geared for the massive umbrella loans to LDC governments that grew up in the 1970s. The government borrowers in LDCs were prepared to place their full faith and credit behind huge loans from the commercial banks that took the loans directly into their asset mix and held them there. In turn, the LDC governments would place funds among their own projects and businesses inside the country. While the largest banks might syndicate some of the loan to smaller banks, all of the loans ended up as assets of the commercial banking system, with about 80 percent going

---

*Since 1982 Citibank, apparently having decided to try to make a virtue of necessity, has expanded its consumer banking facility. Also perhaps relevant is the impact of financial deregulation, which since 1982 has begun to allow banks to participate in the more profitable aspects— such as provision of brokerage services—of dealing with households. Prior to 1982 the banks aimed to make the consumer line of their business profitable with credit cards—prearranged loans up to a specified maximum with hefty interest charges on unpaid balances. Still, the customer that the banks hate most is the one who dutifully pays up every month on the credit card and then tries to negotiate a personal loan at a lower rate than would be paid on the credit card loan.

to the largest like Chase, Bank of America, or Citibank. And as we remember, when any financial trouble threatens, that system provides most of the first-line liquid assets for households and businesses.

The banking revolution of the 1970s concentrated the ownership of claims on LDC governments in the hands of the great metropolitan banks that had arranged the loans. Previous practice would have dispersed ownership over a much broader group of investors in bonds tied to specific projects and not necessarily to governments, and the earlier projects would have had to keep up some sort of track record of efficiency and dependability for more money to be raised later on.

Diversifying Away the Risks?

The bankers who revolutionized their craft during the 1970s thought they knew how to protect themselves from the obvious omission of careful analysis of each project. They would diversify: quite simply, they would not put all their eggs in one basket—have some "chips in every game in town," to use Wriston's colorful but unfortunate phrase, which conjures up the image of Citibank playing the LDC lending game like a gambler in an all-night, high-stakes, poker-players' jamboree. Still, the idea was compelling. If diversification could sufficiently lower the risks, the profits to be reaped from these huge umbrella loans would be enormous.

The key to reducing risk by diversification lies with the fortunes of separate projects being little connected or preferably antithetical, like those of undertakers and life insurance companies. In the case of lending to LDCs, it was seen to operate on two levels. Within each country governments would disperse funds over a wide range of projects that might be expected to do relatively well or poorly at different times so that on average, the whole population, as statisticians say, of projects would yield a fairly stable, hopefully positive return. On the second level, and this probably offered a greater range of diversity, the banks would lend to a group of different countries whose economic fortunes might be expected to move inversely—for instance, oil importers and oil exporters like, say, Brazil and Mexico or Argentina and Venezuela.

The argument for diversification was compelling—but not, it turned out, compelling enough to overcome two very serious problems. First, it overlooked the ability of governments to spend more than they have no matter how much they have. In this case governments proved to be wasteful intermediaries between banks and projects, out of carelessness or incompetence and sometimes out of downright dishonesty. Second, it forgot that the diversification argument relates only to the stability of earnings, not their level. The sad truth is that if a world recession—like the one that began in 1981—unites with waste to produce a widespread negative rate of return, it does not help much to have diversification to thank for its being a *stable* negative return instead of a volatile negative one. The debts still cannot be repaid.

The methods that the banks employed to arrange loans to LDCs, and the size of the loans they arranged, were unique. The heavy reliance placed on so abstract a concept as diversification may have represented the remarkable influence of a relatively new idea in the management of risk, or again it might have proved once more that where there are powerful practical inducements for a course of action, intellectual reasons will surely follow. While the idea of not putting all one's eggs in one basket has been around for a long time— in fact, since its exposition by Daniel Bernoulli in 1730—its full implications were not worked out until after World War II by theorists like Harry Markowitz and James Tobin, now a Nobel laureate. The result was the capital asset pricing model, which could be used to identify optimal combinations of assets based on investor preferences about risk and return. Remarkably enough, there are still numerous areas, including management of foreign exchange risks, where sophisticated practitioners ignore the model's implications.

Nondiversifiable Risks

Even though the LDC lending explosion of the 1970s drew upon some new ideas, they were being applied in territory already familiar to most bankers and they were tactical, not strategic. More loans had to be made. The nature and distribution of those loans were little explored, although they were being contracted in a different world. Governments have been borrowing from their own citizens or from foreign lenders since the twelfth century without creating a threat to the world's financial system as serious as the one that took shape in 1982. In addition to the reasons just discussed, this unpleasant outcome stems also from the unprecedented size of the lending during the 1970s and 1980s and the degree of loans' concentration in the hands of institutions whose responsibilities go beyond international finance to providing households and businesses with the money balances that are at the heart of the world's monetary system.

Governments that borrow from their own citizens create a national debt. When this debt becomes so burdensome that direct taxation to pay for it is no longer politically feasible, governments can, and frequently do, as we have seen, use the inflation tax to lower the real cost of paying it back; this is an elevated way of saying that the state as money monopolist can water down the currency it issues.

LDCs typically have poorly designed systems for direct collection of taxes, much as early governments in the developed countries did. Levying the inflation tax when metal coins were the rule produced some interesting contortions in the design of coins akin to current adaptation in LDCs to an inflationary environment. The profligate Henry VIII debased the English coinage by reducing its precious metal content by 60 percent between 1543 and 1551. Less metal per coin meant more coins with the same face value and more money chasing the same amount of goods. Henry's silver shilling came to be known as the "Red Nosed Harry" because it was so badly debased that the copper showed through at the end of the vain king's nose, which protruded from his full-face portrait on the coin. That coin, which lost 60 percent of its purchasing power over eight years, was the last ever to be minted with a full-face portrait of an English monarch. A strong preference emerged for profiles with the royal ears laid flat against the head. Of course the paper money favored in most LDCs—not to mention industrial countries; wither the silver dollar?—precludes any such embarrassment from troublesome protrusions.

With the appearance of paper money, which became widespread during the eighteenth century, together with the printing press, it was possible for governments to debase money debts even more rapidly. A well-known extreme example came with the German hyperinflation after World War I, which virtually eliminated the purchasing power of the deutschemark. Five-hundred-million-mark notes were common, and by 1923, at the height of the German inflation, waiters changed the prices on menus as patrons gulped down their meals.

While the inflation tax can be highly disruptive, especially when governments use it as heavily as the Germans did after World War I, its affects are largely confined to the country that does the inflating, only operating if a government's borrowing is denominated in a currency whose value it largely controls. Most of the debtor countries in Latin America have already resorted to the inflation tax. Their own citizens suffer heavy losses on local currency loans to their governments, with the proceeds of the inflation tax going to pay back dollar loans. Local inflation just means that more pesos or cruzeros are needed to repay the huge dollar loans that the big banks had so eagerly promoted during better times. But people who are taxed and then receive nothing in return from their governments will eventually rebel, as Brazil and Nigeria—to mention only the most notable examples of that year—learned in 1983.

Extra Risks for LDC Borrowers

Viewed broadly, a government forced into foreign borrowing faces a unique set of circumstances beyond the inability to employ the inflation tax effectively. Questions arise of collateral, of which currency to denominate the loan in, and what interest rate and repayment schedule to set on the loan. The answers that bankers have devised with the governments of developing countries since the early 1970s are in some cases innovative and in other cases downright disastrous.

The banks insisted that LDC loans be denominated in dollars and not in the borrowers' currencies, thereby removing the risk of repudiation by inflation as practiced by Germany with its deutschemark debt after World War I.

The banks even avoided the major risk associated with possible losses tied to dollar inflation. Perhaps they looked at the not-unblemished record on the U.S. government debt in the past, or perhaps they looked ahead to inflation spiraling ever upward. Whatever the reason, they insisted upon floating interest rates on most of their dollar loans to LDCs. If the rate at which money is being lent on the world market—the best working index of inflation's effect on the value of money—rises, so do rates on the LDC loans, which are tied to a lending rate between international banks in London called LIBOR (London, interbank borrowing rate). Since most longer run increases in interest rates result from higher inflation, the banks have largely protected themselves from inflation losses on their LDC lending. The dollar denomination meant they were immune from losses due to LDC inflation. Any attempt to export inflation would just weaken LDC currency against the dollar. The floating rate meant that even inflation in the United States would not harm their position. In the eyes of the bankers, that gave loans to LDCs a leg up even on fixed-interest loans to the U.S. government.

Imprudent Moments

Absence of Collateral on LDC Loans

Even if inflation risks are largely avoided, loans to LDCs are still exposed to risk tied to the inherent economic viability of the projects they are used to finance; and where a loan is made to a state, "economic" includes "political." It does not help much to have inflation protection built into a loan to Argentina if a world recession dries up a large part of its export receipts or if a new government emerges on a platform promising repudiation of oppressive debts to foreigners. For protection against such real risks, banks usually seek some tangible collateral saleable assets that a defaulting borrower has to relinquish if he fails to pay. Mortgage contracts specify that the house or condominium they finance becomes the property of the lender if the borrower defaults.

When governments borrow from foreigners, it is difficult for the foreign lenders to arrange adequate collateral. Some monarchs have offered their crowns and even their own persons, while others offered only a promise to repay, a promise they have cavalierly ignored when it became expedient to do so. This generation of modern governments has offered their foreign creditors remarkably little in the way of collateral beyond the implicit assurance that governments never go bankrupt because they have the power to tax—and, of course, never break their word. But the power to tax can produce burdensome conditions that sometimes result in revolution and overthrow of governments whose successors are not disposed to honor the debts of their predecessors, as we have already seen, many owners of czarist Russian bonds discovered in 1918. Little beyond the power to confiscate airplanes, bank accounts, or other property that is located in a jurisdiction where the courts are sympathetic to the creditor exists in the way of established legal channels or precedent for those injured to seek compensation after a government defaults. What precedent there is still leaves open the possibility that "odious" debts whose proceeds were used by a predecessor government against the interests of its people are not chargeable to the successor.[11]

Risky Business: Lending Before Borrowing

None of these particularly risky aspects of foreign lending were unknown to the bankers who designated their customers, somewhat loftily, as "sovereign" borrowers. But some very hard-driving positive thinkers were running the big banks in the 1970s. When he was made chairman of Citicorp in 1968, Walter Wriston rashly assured shareholders that earnings per share would go up by an incredible 15 percent a year. If he could deliver on such a promise Wriston would gain for Citicorp access to almost limitless capital. If you promise and deliver that kind of performance, investors stand ready to provide a cascade of billions and billions—like the stockbroker at a family gathering who, having doubled a noisy uncle's life savings in a year, finds himself with offers of money from all members of the family, who want their funds doubled. At the same time, a strong performance forces bankers at Chase, Chemical, Manny Hanny, and Morgan's to strive even harder to match it lest they lose out in the intensely competitive atmosphere of big-scale banking.

Delivery on promises of spectacular earnings growth requires very aggressive exploitation of earnings opportunities and a constant pressure to expand volume. For the banks in the 1970s, this meant selling loans. As explained by Al Constanzo, Wriston's vice-chairman at Citibank: "We first go out and make what we consider are credit-worthy loans, and once we have made our loans then we worry about how we fund those loans."[12]

The traditional image of banking has bankers highly solicitous of their depositors' example while being cold and stand-offish with borrowers. But these images are drawn from the small-scale end of the banking business. On the up-scale end of things, up where billions of dollars at a clip are involved, the traditional posturing is turned around. At no time was this clearer than after the first oil crisis. The big problem faced by an OPEC finance minister —if one can imagine such "problems"—was to find someplace to park a billion dollars a month in oil revenues without dropping rates of return through the floor. For these ministers, especially for those innundated during 1974 with the first gusher of petrodollars and seeking short-term, highly liquid outlets for their billions, the big banks were virtually the only game in town. Those in the U.S. Congress who by 1975 asked, they thought sagaciously, what would happen if the Arabs tried to use their oil billions as a political weapon by threatening to withdraw them all overnight from Western banks, would not have worried had they considered what might be done with the money once it was withdrawn. A billion dollars earns $1 million in interest every three days when annual rates of return stand at 12 percent. A credible threat whereby OPEC states pulled $10 billion out of the banks for a month would cost them $100 million in foregone interest. You can buy an awful lot of influence for $100 million a month.

Far from the cringing shopkeeper or small businessman seeking a loan from a skeptical and sometimes sadistic loan officer—remember Banker Potter (Lionel Barrymore) scowling over his glasses at poor George Bailey (Jimmy Stewart), his Adam's apple bobbing up and down in Frank Capra's 1946 film, *It's a Wonderful Life*—the borrower who can absorb hundreds of millions at a time is seen as a prince, especially if the prospective borrower is a government or is able to get a government to "guarantee" the loan. It is not the old British banker's dictum of lending money only to those who do not need to borrow. It is the thought that if we are going to make earnings grow like mad, we have got to up the volume of our business. But that means being prepared to promise prospective depositors of billions that we will pay them millions in interest— more even than our competitors. And to make that promise we have got to have the revenues lined up, which means selling the loans. Since it does not take much more effort to sell $100 million loans than it does to sell $5 million ones, go where the big borrowers are—go to the oil importers. In cases where borrowers were experienced and prudent, as in Western Europe, the mixture of huge oil bills and bankers panting to sell loans was not too volatile. But, to use Anthony Sampson's colorful phrase: "in the more corrupt countries of the Third World like Zaire or Indonesia, selling loans was like offering crates of whiskey to an alcoholic."[13] And for needy countries with poor prospects to repay, selling loans was a form of cruelty, like offering to parents who have not even got the means to buy food easy credit for children's holiday treats and finery.

"I remember how the bankers tried to corner me at conferences to offer me loans," recalls one Latin American who was then Minister of Finance. "They wouldn't leave me alone. If you're trying to balance your budget, it's terribly tempting to borrow money instead of raising taxes, to put off the agony. These gray hairs of mine are because I resisted that temptation."[14]

The bankers' commitment to aggressive growth along with OPEC billions put borrowers in the catbird seat after 1974. The bankers got into the habit of chasing borrowers instead of lenders, and they got careless about distinguishing between the big-country borrowers worth chasing and the corrupt or foolish ones who were willing to go along for the ride, kidding the banks and themselves about "promising" investments for the future. You headed off doubts with optimism and testimonials to "faith in the market system." "There is no power on earth like the power of the free marketplace," Wriston explained, "and governments hate it, because they cannot control it."[15] But it was not always clear that the free marketplace was the only thing the bankers were counting on. The countries to which they were lending billions —Brazil, Mexico, the Philippines, and others—were largely composed of huge government-run corporations. And it is hard to imagine banks' tying up twice their capital in risky loans to LDCs without some thought that the

IMF and the big central banks would be around to smooth things over if the going got too rough.

LDC Loans: Different Views from the Bottom and the Top

The high-sounding rhetoric of bank board chairmen and presidents did not always square too well with the perceptions of the very young men in the trenches who were actually doing much of the lending to LDCs. The following description by S. C. Gwynne, "a banker turned writer," appeared in the September 1983 issue of *Harper's* magazine:

It is 1978. Thanks to the venal, repressive regime of President Ferdinand Marcos of the Philippines, I am safely and happily roosting in one of Manila's best hotels, the Peninsula. I am about to set in motion a peculiar and idiosyncratic process that will result in a $10 million loan to a Philippine construction company, a bedfellow of the Marcos clan—a loan that will soon go sour. I am unaware that any of this is going to happen as I enter the lobby of the Peninsula on my way to dinner, still trying to digest the live octopus that a Taiwanese bank served me last night, and attempting to remember exactly what it was they wanted and why they had gone to so much trouble.

International banking is an interesting business anyway, but what makes it rather more interesting in this case—both to me and to the hapless Ohioans whose money I am selling—is that I am twenty-five years old, with one and a half years of banking experience. I joined the bank as a "credit analyst" on the strength of an MA in English. Because I happened to be fluent in French, I was promoted eleven months later to loan officer and assigned to the French-speaking Arab countries of North Africa, where I made my first international calls. This is my third extended trip, and my territory has quickly expanded. I have visited twenty-eight countries in six months.

I am far from alone in my youth and inexperience. The world of international banking is now full of aggressive, bright, but hopelessly inexperienced lenders in their mid-twenties. They travel the world like itinerant brushmen, *filling loan quotas,* peddling financial wares, and living high on the hog. Their bosses are often bright but hopelessly inexperienced twenty-nine-year-old vice presidents with wardrobes from Brooks Brothers, MBAs from Wharton or Stanford, and so little credit training they would have trouble with a simple retail installment loan. *Their* bosses, sitting on the senior loan committee, are pragmatic, nuts-and-bolts bankers whose grasp of local banking is often profound, the product of twenty or thirty years of experience. But the senior bankers are fish out of water when it comes to international lending. Many of them never wanted to lend overseas in the first place, but were forced into it by the internationalization of American commerce; as their local clientele expanded into foreign trade, they had no choice but to follow them or lose the business to the money-center banks. So they uneasily supervise their underlings, who are the hustlers of the world financial system, the tireless pitchmen who drum up the sort of loans to Poland, Mexico, and Brazil that have threatened the stability of the system they want to promote.[16]

Young Gwynne's Philippine loan eventually was made after being partially guaranteed by the Philippines' largest bank, which was "handing [guarantees] out like free samples." A year and a half later interest and principal payments on the loan simply stopped coming. The loan was rescheduled, but as of summer 1983 the bank had received only a fraction of the money owed. The guarantee was never called. By October of 1983, in the wake of political unrest over the August 21 assassination of Benigno Aquino, Jr., the Philippines declared a ninety-day moratorium on repayment of its $18 billion in external debts. Its central bank governor, Jaime Laya, described the problem as one with political rather than economic roots. Subsequently the Philippines requested a "further breathing spell" until 1985 on its debt repayments.[17] Meanwhile Gwynne had, in 1980, left the bank holding the Philippine loan for a job —at twice the salary—with a big West Coast bank. After a brief stint at that job he turned to writing.

Viewed from Gwynne's perspective—close to the bottom of the ranks of international bankers—as well as from Walter Wriston's perspective—at the pinnacle—there was a way, perhaps unimagined by most outsiders to banking, in which the lending game was intensely competitive. You had to sell more loans than your competition. And by 1980, after the Europeans and Japanese had gotten in on the lending boom act, the competition was stiff indeed, pushing spreads over the cost of funds down to less than a percentage point. Some reflective types asked whether Zaire or the Philippines ought really to be assessed risk premiums of only half a percentage point over AT&T, but there you were, the competition to lend—that was the way to grow and increase earnings—kept the spreads low and, besides, countries, "sovereign borrowers," do not default. LDC lending was where the action was. Even during the 1981 recession there was plenty of it, especially in the short-term and overnight money markets where, at least until August of 1982, Mexico and Brazil were rolling over billions to pay finance charges on longer term debts.

The End of the LDC Lending Boom: Some Frightening Mathematics

The banks' attitude toward the LDC lending boom changed radically after the Mexicans dropped their debt bombshell on Washington, D.C., right in the middle of all those well-deserved bankers' August vacations. As a result, the kick-off of the fall season a few weeks later, the IMF–World Bank meetings in Toronto over the 1982 Labor Day weekend, which are usually a blast wherever they are held, were a pretty gloomy affair: "It was the Titanic. We were just rearranging the deck-chairs," said Wriston.[18] Unflappable as ever, the Citibank chairman wrote his now-famous article in the *New York Times* after the meetings "to put an idea into the marketplace." Now "a great believer in the competition of ideas"—if not in letting the "marketplace" take care of shoring up Citibank's tapped-out clients in Mexico, Brazil, and elsewhere— Wriston wrote that governments never repaid debts anyway and could not. What mattered was not "ability to repay" but rather "access to the market-place."[19] This meant that as long as loans were available to provide mega-borrowers with the billions they needed to pay the interest on their loans, no one really needed to worry about defaults or arrearages. The following day in his "Economic Scene" column Leonard Silk quoted former U.S. Treasury Under Secretary Robert V. Roosa, now a partner in Brown Brothers Harri-man, who described Wriston's article as "soporific" and "just plain cotton candy."[20]

Wriston was right in a way, but only in a very special way. Suppose we take Brazil, with debts totaling $90 billion, carrying an average interest rate of 12 percent—or $10.8 billion per year. Leave aside for now repayments of princi-ple. If Brazil borrows the $10.8 billion to repay interest, then by year end its debts total $90 billion plus $10.8 billion, or $100.8 billion. Twelve percent interest payments then require borrowing another $12.1 billion, and total debt rises to $112.9 billion, and so on. At the end of ten years, Brazil's $90 billion debt becomes $90 billion times $(1.12)^{10}$, or $279.5 billion. The miracle of compound interest works against borrowers just as hard as it works for lenders. Of course, the only case in which "access to the marketplace" works to rationalize borrowing to repay interest is where the means to repay the loan also grows at 12 percent a year so that the ratio of debt to, say, GNP stays constant. It is necessary to squeeze inflation distortions out of the calculations —GNP could grow at 200 percent a year measured in cruzeiros with 200 percent inflation and the real ability to pay dollar interest charges would be unaffected—in order to be sure to get it right. The "real" interest burden in a 12 percent nominal interest rate over the next ten years would be about 7 percent, given an average 5 percent inflation rate over the period. If Brazil's real GNP growth rate—or alternatively, as suggested by some, the real growth of Brazil's exports—were 7 percent for the next ten years, the scheme of borrowing to pay interest would just keep that nation's credit-worthiness constant at today's questionable level. Even Brazil's gold mines can't come close to supplying the roughly $1 billion per month needed to service her debts.

Imprudent Moments

Under these circumstances, any improvement in Brazil's credit-worthiness would require real growth in excess of 7 percent, real interest costs of below 7 percent, or simply borrowing of less than interest charges. The latter means getting the money from somewhere else. Externally, if more borrowing is to be avoided, it would have to be a gift. Barring huge ongoing gifts of billions of dollars, the only other place is from restriction of domestic consumption to free up internal saving. This is where the politically destabilizing austerity programs come in. They are very painful, as a look at reports on conditions in Brazil and Mexico during 1983 will show. As such they must be temporary—aimed at squeezing an extra $4 or $5 billion out of the home economy for one or two years at most. If Brazil could reduce additional borrowing to $5 billion for two years, then at the end of that time debts would total $100.3 billion instead of the $112.9 billion with $10.8 billion a year in borrowing. Then, even returning to the original scheme of borrowing to pay interest of 12 percent, total debts would reach $247.5 billion in ten years instead of the $279.5 billion under the original plan. A reduction of $11.6 billion in external borrowing over the next two years would mean $32 billion less in debt at the end of ten years. Austerity is risky and painful, but every dollar saved now is worth nearly $3 at the end of a decade under typical, foreseeable conditions.

Furthermore, given the smaller $247.5 billion debt ten years in the future, constancy of credit-worthiness would require a lower annual GNP or export growth of 10.6 percent—5.6 in real terms with 5 percent inflation. A reduction in the required growth rate of 1.4 percentage points brings it into a range far more feasible for Brazil, which has demonstrated—at least in the past—a long-run ability to grow at an average real rate of 6 percent.

Bank Cooperation Replaces Competition

Wriston's 1982 reference to "access to the marketplace" did not mean access to a competitive marketplace wherein the banks were falling all over each other "competing" to sell loans to developing countries. It meant instead a radical transformation from intense loan-selling competition among banks in pursuit of profits to a system of collective lending that involved governments, the IMF, the BIS, the World Bank, and the banks in a sometimes-desperate catch-up effort to protect their past commitments. Default became unthinkable since it would more than eradicate the shareholder's equity at the world's largest banks and do God knows what to the world economy. Arrearages on debt service payments became easy enough to imagine—with their possibly devastating impact on earnings and then on the ability to raise capital and grow!— to prompt a downright scramble to make damn sure they did not occur. The last thing banks wanted was an official triggering of a clause on arrearages, with its devastating impact on earnings. To avoid it, and worse, they lengthened sixty-day arrearage clauses to the maximum ninety days, swallowed hard and ponied up billions when the IMF "bailed them in" on Mexico and Brazil, and put together teams of bankers to raise the banks' share of rescue packages especially from recalcitrant regionals and some "free riding" British and Japanese banks.

The teams were composed of fiercely competitive bankers who just a year before would never have dreamed of sitting down together to put together a loan for Mexico or Brazil but would have been lobbying intensely in Mexico City or Rio to grab the loan for their own banks. Men like Morgan's Antonio Gebauer, described as an "ambitious son of German immigrants who settled in Venezuela"[21] and Bank of America's Rick Bloom "with a penchant for allegorical anecdotes"[22] had to get used to cooperation instead of competition. Since the highly competitive nature of international lending was exactly what had attracted hard-driving, aggressive personalities like Bloom and Gebauer —not to mention Wriston himself—it could not have been easy to start operating like a lender's cooperative, "taking from each [lender] according to his ability" and "giving to each [borrower] according to his needs." Wriston's "access to the marketplace" was just one attempt—of many—at a euphemism to make a little less painful the distasteful process of reining in all the competitive juices that had been flowing for almost a decade.

Locking the Wheels

What remains is to determine what force on earth could bring so sudden a stop to the music that both borrowers and lenders had been so pleased to hear and dance to with such abandon from 1974 to 1981. In two words, it was Paul Volcker. Again, Wriston is the spokesman who characterizes a transition from the banker's perspective. Asked in a 1978 interview whether he would welcome any restrictions on any U.S. lenders or borrowers, Wriston replied: "I just want the opposite. Let us have their freedoms." "Which ones?" asked the interviewer. Wriston replied: "I believe the most important thing the bank will have to deal with over the next 10 years is not money policy, because the options are limited and there isn't much elbow room. [Rather] it's the revolution in the financial business of America."[23]

Five years later, asked the inevitable "what's gone wrong now" question in another interview just before the September annual meetings of the IMF and World Bank, Wriston replied: "We're beat upon the fact that we have imprudent moments. But I don't know anyone that knew Volcker was going to lock the wheels of the world."[24]

No one did know, save one: Paul Volcker. And he told everyone in a rare radio message broadcast on a Saturday afternoon early in October of 1979. But for a while no one listened.

PART IV

Lenders of Last Resort

Chapter 8

Angels, Guardian and Avenging

> In the final analysis, the lender of last resort must make decisions.
>
> CHARLES P. KINDLEBERGER,
> *Manias, Panics and Crashes*

THERE ARE two legitimate reasons to borrow—"legitimate" in the sense that any debts contracted can be serviced and, ultimately, repaid. They are investment financing and adjustment financing. Failures in connection with investment financing are fundamental and result in solvency (negative net worth) problems for lenders, which they may try to pass off as liquidity (cash flow) problems in order, inappropriately, to enlist the aid of a lender of last resort. Failures of adjustment financing may result in liquidity problems that, appropriately, a lender of last resort should prevent from degenerating into solvency problems by providing bridging loans. If, alternatively, the lender of last resort errs in labeling a liquidity crisis a solvency crisis, the result may be an unnecessary financial collapse.

In view of these distinctions, it is important for lenders and borrowers to be clear, prior to consummation of their deal, about which type of transaction they are entering into. If they fail to make such a distinction, then should problems of repayment arise after the fact, the lender of last resort must make the distinction in order to determine whether intervention is appropriate (adjustment) or inappropriate (investment).

Grossly characterized, the financial crises of the early 1930s—especially in the United States—resulted from difficulties with adjustment loans and attendant liquidity problems that were allowed, by virtue of the Federal Reserve's error in treating them as solvency problems, to degenerate into solvency problems. By contrast, the debt crisis of the 1980s has resulted largely from difficul-

157

ties with investment loans to governments—for which the investment/adjustment distinction admittedly is harder to make than for businesses—that after 1980 were incorrectly treated by private banks and some governments and international agencies as adjustment problems. The result has been a large volume of short-term adjustment loans piled on top of investment loans that the lenders—incorrectly—have interpreted as appropriate candidates for aid from a lender of last resort.

Investment Financing and Adjustment Financing Examined

The purpose of investment financing is to provide the borrower with funds necessary to undertake a prospectively profitable investment, such as building a factory to produce energy-efficient automobiles after a quadrupling of petroleum prices that is expected to be permanent or, on a larger scale, building a large hydroelectric project in a country expected to be industrializing rapidly following, say, a strong and likely permanent expansion in demand for the country's manufactured output.

The purpose of adjustment financing—usually undertaken for the shorter term—is to finance expenditure during an interruption of revenue or a rise in expenditure that is reasonably expected to be temporary and reversible. The "reversible" part is required to provide the wherewithal to pay off the debt that comes due once the "temporary" drop in revenue or rise in expenditure is replaced by a roughly equivalent temporary rise in revenues or drop in expenditure. The classic case surrounding the latter type of borrowing is the down phase of the business cycle. It is temporary and very likely to be followed by an up phase—hopefully of roughly equal or greater intensity and duration—during which the excess of revenues over expenditures swells sufficiently to pay off debts.

The Lender's Problem Is the Borrower's Problem

The problem for lenders as they face borrowers in search of funding is always the same. They must evaluate the accuracy of the prospective borrowers view of the world. Since every borrower has on hand for use in dealings with rational lenders some description of a future world wherein he readily services debt and eventually either repays or rolls over (reborrows) debts on attractive terms for himself and lenders, it is left up to lenders to evaluate the plausibility of what are necessarily predictions by borrowers. In the case of longer term investment projects such as a hydroelectric project, the lender must decide on the suitable scale of the project, the ability of the borrower to carry it out, and, most important, the likelihood that predictions which make the project prospectively attractive will be realized. For shorter term adjustment loans the lender has the very difficult problem of analyzing—before the outcome is known—whether a "temporary" drop in revenues relative to desired expenditure will in fact turn out to be temporary and, beyond that, whether it will be replaced by a timely rise in revenues relative to desired expenditure required to pay off the accumulated debt. If the change in the relationship between revenues and expenditures is not temporary, borrowing is the wrong response. Rather, some adjustment is required whereby in the event of a perceived need to borrow, expenditures are cut, revenues are raised, or some combination of the two brings them back into balance. Usually expenditure cuts can be effected more quickly than increases in revenues that require producing and selling more goods, for if it were possible a prospective borrower would already have produced and sold more, as means to reduce required borrowing or to convince prospective lenders of a greater ability to carry debt. In short, borrowers whose expenditures are expected to exceed revenues for the foreseeable future are living beyond their means. Less spending, not more borrowing, is what is called for.

Going International

When borrowing and lending across national borders is considered, the process is described using some different terminology. Revenues become exports and receipt of interest from abroad, while expenditures become imports and payment of interest to foreign lenders. The difference between revenues and expenditures so defined becomes the current account surplus (positive) or deficit (negative). As with any household, a nation's difference between revenue and expenditure during any time period results in borrowing, given a current account deficit, or lending, given a current account surplus. Borrowing from other nations is labeled a capital inflow, while lending is deemed a capital outflow. The difference between capital inflows and outflows measures the capital account on net capital flows. A nation's balance of payments is the sum of its current and capital accounts. If the two accounts do not sum to zero, nations—like households—have only one real alternative. They must either draw down reserves if the balance is negative, say with capital inflows falling short of a negative current account balance or—a much easier position to be in—they must accumulate reserves if the balance is positive. Exchange rate adjustments that nations may employ are no more than ways to adjust expenditure by lowering the value of domestic financial assets with a devaluation or enhancing their value with a revaluation.

Why a Debt Crisis?

The initial phase of the world debt crisis of the 1980s is a result of what—after the fact, at least—has been excessive lending by banks, international agencies, and governments of developed countries largely to governments of developing countries. The process began quite legitimately in terms of the normal long-term criteria for lending and borrowing. LDCs like Brazil, Mexico, Korea, the Philippines, and numerous others wishing to grow and expand capacity naturally wanted to spend in excess of their relatively low incomes in the immediate postwar period. In many cases, prospects for sale of output from their targeted export industries were good. Steel produced in Korea with modern equipment together with relatively cheap and remarkably hard-working labor would be very competitive in world markets and therefore would serve as a reliable and growing source of export revenue, provided that Korea could obtain the financing needed to build its own mills. Provided further that Korea's steel mills could be expected to keep pushing out rolled sheet and heavy girders in quantities that expanded at more than 5 or 6 percent a year, the usual criteria for successful debt servicing could be met. If unforeseen interruptions along the way—like a quadrupling of oil prices or a wave of protectionism in developed countries—meant temporary reversals, additional, temporary, short-term financing combined with some belt-tightening adjustments could probably see them through.

But what about a Brazil? With half a century of rapid growth behind it, it became ever more inward looking as the twentieth century passed by. It seemed to seek more and more to imitate its wealthy cousin to the north, which a century earlier had developed spectacularly with some attention to trade with other nations but essentially as a result of exploitation of its own tremendous resources, the quest for which resulted in an ever-expanding home market. Brazil looked outside itself primarily for the capital needed to build its factories, roads, and hydroelectric dams, just as the United States had done. Perhaps seeing something of its earlier self, perhaps more comfortable investing in a resource-rich country in its own hemisphere, the United States poured billions into Brazil. After 1973 most of the billions were poured in by U.S. banks, which, for a combination of reasons explained in chapter 7, were poised for an explosion of foreign lending just as the oil shock came along and supplied a classic rationale for a very selective increase in short-term adjustment lending.

But as they expanded their loans to Brazil and most of the developing world at an unprecedented rate after 1973, U.S. banks confused a fundamental distinction between long-term investment and short-run adjustment assistance. Their confusion was compounded by a profound error of economic policy in the developed countries. After what amounted to a massive structural shock that sharply reduced the ability of the world's economic machinery to produce food, automobiles, steel, and almost every conceivable commodity with existing blends of labor and capital, governments of the industrial countries, led by the pathetic U.S. locomotive, tried some old-fashioned pump-priming to push up demand for goods that just were not there. The result was inflation—far worse than the fairly moderate dose of demand that was administered would normally have produced. The reason for the unusually virulent epidemic of inflation during 1977–79 was that a sharp change—less use of petroleum energy and other raw materials—in the requisite mix of inputs had not been fully effected and in some cases—especially in the energy-guzzling United States—thanks to artificially low energy prices, had scarcely begun.

The result of the unexpectedly sharp burst of inflation in the late 1970s was to validate what ought to have been excessive bank lending in response to the first oil crisis. Higher commodity prices and in some cases low inflation-adjusted interest rates on adjustment loans made the Brazils and even the Philippines of the world look like solid prospects when the second oil shock hit. But by then the industrial countries had discovered that demand does not create its own supply while also realizing that another round of trying to prove true such a preposterous idea would bring many to the brink of a ruinous hyperinflation. At the same time remarkable progress—much of it still unappreciated as late as 1981—had been made in adjusting the input mix to reflect new scarcities and more costly energy. Lighter alloys are a very effective substitute for energy inputs into moving machinery. As a result of the unperceived adjustment, the negative response to tighter money after the second oil crisis was far sharper than expected. Prices—especially those of energy-related raw materials, which had soared during the late seventies—fell far more sharply than anyone had guessed they would. Inflation-adjusted interest rates shot up from 1978 levels close to zero to lofty 1982–83 levels of over 10 percent at times—incredible in view of historical norms of between 2 and 3 percent.

Confusion Between Investment and Adjustment Loans

The negative shocks after 1980 sent LDCs rushing to their banks for short-run adjustment loans. In 1981 the banks swallowed hard and pretended they were seeing more of the temporary and reversible disturbances that would justify some short-term loans. As a hedge they kept the term short on much of the new lending—six months to a year—expecting to review an improved situation after that interval. But by 1982, when the time came to roll over the loans, even the rose-tinted glasses the bankers and their clients had donned a year earlier let show through some very dreary prospects. The rollovers stopped, first in 1982 for Mexico, then in 1983 for Brazil and a raft of other borrowers. The debt crisis was on, and the remaining questions centered on what if any lender of last resort might emerge and, unavoidably, on who was going to absorb huge losses in a world so unlike the one that had been foreseen only a few years earlier.

Moral Hazard?

In the minds of some skeptical souls, one way to explain the imprudent rush by the banks to lend billions to tottering LDCs is to postulate that, in spite of their espousal of free enterprise, the bankers had it in mind that if things really got rough an angel in the form of the much-revered lender of last resort would appear to bail them out. This is the "moral hazard" problem common to all insurance schemes. After all, some well-placed expressions of concern by New York bankers—carefully coached by Felix Rohatyn, a partner of Lazard's investment banking house, to worry in unison at the fall 1974 IMF meeting about the possible heavy damage to the financial system of the whole world should New York be forced to default on its bonds—brought forth from a somewhat reluctant, though pliable Gerald Ford an agreement to ask Congress for a $2.3 billion loan to tide New York over its problems. The day was saved and, some say, the precedent created whereby bankers, who might have begun to worry earlier about the billions flowing abroad to LDCs, drifted off each night into peaceful sleep with visions of angels beating down any lurking demons of default.

The international lenders of last resort manifest themselves usually in two forms: the International Monetary Fund and the Bank for International Settlements, each backed of course by the massive resources of powerful governments concerned on behalf of their powerful banks. As the most powerful representative of the richest government, the U.S. Federal Reserve System always appeared as the most vivid image standing behind the two primary angels of the world's system of finance.

The effectiveness of these angels and their allies in warding off the demons of default depend on two things: the inclination and the capacity to do so. As for inclination, no lender of last resort can reveal in advance a willingness to rescue any overlent bank, lest it be placed in a position of encouraging imprudent, excessive exposure for the sake of higher earnings. This is the "moral hazard" problem applied to banking. More generally, it raises the question of how an insurer prevents insured parties from foregoing normal, prudent behavior; how to keep the homeowner with theft insurance from carelessly leaving his house unlocked. Deductibles, sentimental attachment to irreplaceable items, and the "hassle" of arranging for payment and replacement are effective deterrents for individuals, but these may not be so effective as inducements for cautious behavior by large institutions such as banks.

Borrowers: Servants or Masters to Lenders?

Just as the borrower whose loans represent a significant chunk of a bank's assets gains leverage over his creditor—seemingly to contravene the biblical proverb that "the borrower is servant to the lender"—so banks whose assets represent backing for a big chunk of the money supply and other holdings of households and businesses gain some leverage over their creditor, the lender of last resort at the central bank, as soon as the value of their collective assets is threatened. The leverage comes from the fact that banks' equity cushions —their net worth—is only 4 or 5 percent of their total loans, while earnings average only about one-half to three-quarters of 1 percent of total loans. A loan loss equal to 1 percent of its portfolio for a huge bank with $100 billion in assets —$1 billion—can wipe out up to two years' earnings. In short, banks are highly leveraged.

When big banks try to convert their highly leveraged state into heavy leverage on lenders of last resort, they raise immeasurably the level of tension prevailing in world financial markets. Attention is unavoidably focused on the crisis that has brought the banks to seek a "bailout." Bankers put their case for help in terms of the "dire" consequences—which "no one wants to see happen"—for the financial system should they fail, usually neglecting to point out that the biggest losers would be shareholders in the bank, not its depositors who can be and in many cases are protected by deposit insurance. But such scare tactics are effective—they worked in 1974 with the New York City crisis and many bankers obviously expect them to work in the 1980s with the world debt crisis.

For their part, the lenders of last resort—which we shall consider at greater length in the next chapter—must not appear to reward imprudence. This means assessing their charges some of the costs of their carelessness—discounting freely but at a stiff penalty rate in Bagehot's time or "bailing them in," to use the contemporary phrase of IMF Managing Director Jacques de Larosière—but not so much so as to impair their solvency, which in the case of particularly imprudent behavior, such as Citibank's $5.0 billion in loans to Brazil—its entire net worth—unavoidably means a bailout of a magnitude that rises with the degree of imprudence. The alternative—the logical course of letting losses rise in proportion to imprudence—means eschewing the lender-of-last-resort objective of saving the system by saving its worst actors. The inclination of lenders of last resort to don their wings and pitch in to save the system depends on their assessment of the net effect of all these discordant forces.

Insuring Depositors, Not Shareholders

Quite apart from the inclination of the lenders of last resort to rescue the system is their ability to do so. And the definition of "rescuing the system" depends crucially on whether one is thinking of rescuing bank depositors (appropriate) or bank shareholders (questionable). Bank depositors have contracts with banks that involve promises to pay "on demand" (demand deposits) or "within a specified time" (time deposits) one dollar plus any accrued interest for every dollar deposited. Bank shareholders, in contrast, have no such guarantee. Their capital is at risk, just like that of the shareholders in any corporation. In theory, if a $100 billion bank with shareholder's equity (net assets) of $5 billion loses that amount or more, it comes out of the shareholder's hides. In practice, as grateful owners of Lockheed and Chrysler shares (to mention only two recent examples) know, governments can be convinced to bail out shareholders, essentially based on Charles Erwin Wilson's famous rationale, provided to the Senate Armed Forces Committee in 1952, that "what's good for General Motors [or Chrysler, or Citibank, or the Bank of America, for God's sake] is good for the country."

With these visions of our lenders of last resort and sugarplums, like a nice juicy bailout, dancing in our heads, it is interesting to examine our two reluctant saviors who must struggle with the forces—sometimes pushing them toward the role of avenging angel, while at other times pushing them to be merciful angels. However they decide to act, our two angels have not anywhere near the capacity to make everything better. The Bank for International Settlements can only commit, for short periods of time, a few billions in "bridging loans" provided by the central banks, which are either members or supporters. (The U.S. Federal Reserve is not a member.) After the November 1983 passage of the bill to include $8.4 billion in U.S. support for the International Monetary Fund and its arm, the General Agreement to Borrow, the Fund added about $30 billion in usable currencies to its resources over the next three years. While a huge sum, it falls far short of sums widely acknowledged as being necessary to keep some $300 billion in LDC bank loans afloat for the rest of the decade. The banks themselves must put in tens of billions of dollars more, and in so doing they face the dilemma of when, if ever, to stop throwing good money after bad.

Chapter 9

From on High

What was Bretton Woods about if it was not the creation of a world in which countries did not close their eyes to the repercussions of their actions on others?

AUSTIN ROBINSON, "A Personal View,"
Essays on John Maynard Keynes

There never was any gentleman's agreement of Basel, but at Basel there are only gentlemen.
WILFRID BAUMGARTNER, French Finance Minister, statement at a Vienna press conference, September 1961

THE International Monetary Fund, together with its sister institution, the World Bank, was born in July of 1944 at the little resort village of Bretton Woods, New Hampshire. It opened for business in Washington, D.C., in 1947, months after a March 1946 Inaugural Meeting of the Boards of Governors of the IMF and World Bank at Savannah, Georgia. In the eyes of its intellectual founders, John Maynard Keynes and Harry Dexter White, an Assistant Secretary of the U.S. Treasury, the IMF was meant to operate as a true world central bank that could function as an international lender of last resort or at least help to avoid the need for such action by forcing timely adjustments to payments imbalances. The World Bank or, more appropriately, the International Bank for Reconstruction and Development (IBRD), was to provide longer term financial assistance to nations, particularly poorer ones, thereby concentrating on economic development, while the IMF was to serve at the center of a new international monetary system with stable exchange rates between national currencies serving to encourage growth of free trade and capital flows. Despite its imposing title of World Bank, the IBRD was designed to have a good deal more to do with the developing countries of the world and their long-term projects to aid economic growth, while the IMF was to be an organization that maintained the payments mechanism among the advanced countries. A system with the IMF at its center was meant to replace the International Gold Standard, which had collapsed in 1931, and to avoid the exchange-rate volatility symptomatic of inward-looking policies followed during the 1930s by the major industrial powers.

As the postwar period unfolded, the IMF was pushed further and further out of the key role envisioned for it by its founders. Like other international organizations, the Fund found that strong nationalism remained alive and well even after the second devastating world conflict in a little over a generation. More and more it was pushed out of an effective role in designing adjustment measures for the major countries pursuant to exchange-rate stability and into a role of expediting short-term adjustment assistance for the stronger developing countries that had graduated from the need for development loans from the World Bank. Countries like Brazil, Korea, Taiwan, Venezuela, Mexico, and Argentina were getting more and more of their funding from commercial banks, but they still needed IMF financing to cushion them from short-term balance-of-payments problems arising from volatile prices of their commodity exports.

It is important to understand the evolution of the IMF role in its dealings with developed and developing countries during the postwar period in order to see what lay behind its dramatic emergence in 1982 as an intermediary between the banks and heavy borrowers in the Third World. The forces determining the evolution began to build during the turbulent interwar period.

The IMF's Interwar Roots

The abortive effort forged by England's Sir Montagu Norman and the United States' Benjamin Strong that had pushed prices of U.S. securities to dizzy heights in 1929, before the October collapse and subsequent slide into depression, had left the United States looking inward by the time Franklin Delano Roosevelt came to office in 1933. The debut of the United States as England's replacement in the role of world economic leader had ended inauspiciously; Europe was left unimpressed. Repeal of Prohibition occupied more U.S. attention in 1933 than did the state of the international economy. Benito Mussolini, perhaps less reserved than other European leaders, declared in 1933: "I can sum up the United States in two words: Prohibition and Lindbergh!"[1] Bitter disappointment with the promise of emerging U.S. hegemony in the 1920s— perhaps epitomized by Lindbergh's triumphant transatlantic flight in 1927—

had reached a point in the early 1930s where Lindbergh's name was cruelly invoked by a European dictator to characterize the United States as a land of gangsters and kidnappers.

Il Duce had a sharp tongue given to unfair characterizations that served his purposes by telling people who did not know any better exactly what they wanted to hear. But sometimes it served his purposes to speak the truth. Asked his opinion of U.S. foreign policy in 1933, Il Duce said: "America has no policy."[2] The new U.S. President Roosevelt did little in that year to contradict this blunt view, silencing official voices in favor of U.S. participation in the League of Nations and torpedoing the London International Monetary and Economic Conference in June of 1933—remember "First things first."

Ironically it was John Maynard Keynes who, almost alone among economists, declared FDR's negativism on the London Conference to be "magnificently right."[3] Keynes was still smarting over Britain's abortive 1925 return to the gold standard. Undertaken over his objections by the Chancellor of the Exchequor, Winston Churchill, it had forced the disastrous U.S. easy-money policies that were required to keep sterling tied to gold. An obvious need for a devaluation of sterling against the dollar—which would have resulted from tighter money in the United States—would have signaled all too clearly sterling's overvaluation in terms of gold. The whole thing had collapsed when, after the failure of Kredit Anstalt in May 1931, runs on Britain's gold by the French and Dutch forced Britain off gold in September 1931. Keynes preferred managed currencies to the gold standard and Roosevelt, by turning inward in 1933 to concentrate on U.S. problems at home, assured that no new gold standard would soon be established.

There did emerge out of the cooperative spirit of the 1920s one institution that survived the collapse of cooperation and financial markets in the early 1930s. The Bank for International Settlements was conceived during the late twenties to depoliticize financial matters concerned with German reparations. The BIS held its first meeting in May of 1931. Twenty governors of central banks, the greatest number ever to have appeared at a single gathering, attended in Basel, home of the BIS. Fred Hirsch aptly described this first gathering as "a meeting of sovereigns whose empires were about to erupt."[4]

The BIS is an institution distinguished by its European character, its durability in the face of adversity, and its membership composed strictly of central bankers. The U.S. central bank was not represented at its first meeting, and the United States has never officially joined the BIS, although a member of the Federal Reserve Board of Governors—currently Henry Wallich, a cigar smoker in the best clubby tradition of central bankers—has regularly attended its monthly meeting in Basel since the early 1960s. Even though within three months members of the fledgling BIS fraternity of bankers were severely at odds, the regular monthly meetings continued and were well attended. The British abandonment of gold in September of 1931 cost the Bank of France $90 million—five times its capital; the Netherlands Bank suffered even more heavily, thanks to its deferral of withdrawals from the Bank of England based on assurance from London. The charge of "breach of faith" cut deeply at the Bank of England—turning devaluation into a personal, moral issue for its governors and likely accounting for the painful delay in the devaluation of sterling thirty-six years later, in 1967. The Dutch, for their efforts, were subjected to a nasty little ditty:

In matters of sterling the fault of the Dutch
Is caring too little and trusting too much.[5]

Despite these heavy strains, the civilized coherence at the BIS was maintained. It was really like a club for central bankers and has often been characterized as such. Beleagured heads of central banks could meet there—in privacy, or some would say secrecy, since until 1977 the entrance to the BIS was an unmarked doorway next to Frey's Chocolate Shop across from Basel's railway station—and count on sympathetic ears to listen to complaints about how hard it was to maintain monetary order in the face of heavy political pressures from national treasuries to print more money. Three decades later, when the Bretton Woods system, with the IMF as its nominal head, began to show signs of coming apart during the gold bubble of 1960, it was at the BIS in Basel, not at the IMF in Washington, where American central bankers joined with their European cousins to devise the gold pool, currency swaps, and general arrangements to borrow that were to prop up the system of fixed exchange rates for another decade—almost like a life support system that keeps the patient alive long after he would have died on his own.

American Views of the IMF Prevail over British Views: A Weaker IMF Is Created

Somewhat ironically, in view of Keynes's vision of the IMF as a true world central bank able to transcend by responsible control of world liquidity the need to tie money issues to gold—"the barbaric relic"—much of the effort by the major powers during the 1960s was devoted to propping up the dollar's shaky link to gold. This came about largely because the Bretton Woods system, contrary to its original design, emerged after 1946 as a system that thrust most of the burden of adjusting to international payments imbalances onto deficit countries. Keynes had wanted to have adjustment to imbalances shared equally by deficit and surplus countries alike. The asymmetric adjustment burden on deficit countries that emerged was most unfortunate because it reasserted the mercantilist tendency—never very far in the background—that balance-of-payments surpluses are somehow "good" while deficits are "bad."

Keynes's device for putting adjustment pressure on surplus countries was the "scarce currency clause" in the Bretton Woods Agreement. Basically, the way Keynes's concept of an international clearing union worked had IMF members paying in gold and their own currencies to a pool of reserves usable by deficit countries conditional on their taking measures to adjust. The deficit countries could draw from the pool the currency of the country against which they were running a payments imbalance. As a result, the currency of countries persistently in surplus would become "scarce" in the pool. Once the pool's holdings of a country's currency fell below a predetermined level, the scarce currency clause triggered a requirement for such persistent-surplus countries to take measures to adjust domestic policies—say, toward easier monetary policies.

This arrangement could impart an inflationary bias to the world financial system unless careful attention were paid to maintaining overall an appropriate world supply of money. Keynes was not deterred by this fear for two reasons. First, the bias in the 1930s, after the liquidity surge of the late 1920s designed to prop up sterling, had been on the deflationary side. Second, Keynes had envisioned the IMF as a true world central bank, which meant it would be able to control the overall supply of world money.

The Americans, represented by the U.S. Treasury and Harry Dexter White, were suspicious of Keynes's plan as a back-door way of arranging easy loans for Britain, which by 1945 was badly in debt. The easygoing British attitude about "overdrafts" (negative bank account balances) was not well accepted in the United States until the advent of credit cards in the 1960s, which permitted millions of Americans—by then transformed to a population in which fewer and fewer people remembered the Great Depression—regularly to spend well in excess of their current bank balances.

In a way the Keynes plan was idealistic and advanced, based as it was on the notion that, after the painful lessons of the 1920s and 1930s, the world's financial community ought to realize that enlightened self-interest combined with a sophisticated understanding of how the world worked ought to dictate control of money instead of a mechanical link to gold. The ill-fated British return to gold in 1925 had not saved the interwar world from the calamity of depression and—it may have been thought—might even have had something to do with causing the Great Depression. Keynes's strong opposition to gold persisted throughout the 1930s. When in 1940 he heard some British ridicule of messages broadcast by the Nazi German banker Dr. Funk about a postwar New Order that included abolishing the monetary role of gold he remarked: "About three quarters of the German broadcasts would be quite excellent if the name of Great Britain were substituted for Germany or the Axis."[6]

Keynes's sophisticated and cosmopolitan views on money were not shared by the Americans, least of all the U.S. Treasury. American distrust of banks ran deep, dating back to the Jeffersonian dislike of credit, strongly expressed in opposition to Alexander Hamilton's 1790 establishment of the Bank of the United States, which he modeled on the Bank of England. In his landmark book *Banks and Politics in America*, Bray Hammond ties American dislike of banks—especially central ones—to regionalism prevalent in the nineteenth century. As the West and the South developed, there emerged an aversion to the Bank of the United States motivated by "complex psychological and political considerations—including past distress and present dependence— while its New York enemies were moved much more simply by covetousness and rivalry."[7] Negative images of the greedy, unfeeling banks abound in American literature. John Steinbeck's *Grapes of Wrath* has bulldozers materialize along with gun-toting sheriff's deputies as the agents of the big banks driving dustbowl farmers off farms they thought they had come to call their own because they—not the bankers—had worked the land that only a piece of paper said was owned by the bank. Banks represent too abstract a concept of ownership for Americans, who remember how the country was expanded and developed by men and women—homesteaders—who were given land in exchange for occupying and cultivating it. Banks, like insurance companies, always seemed most agreeable and friendly when you least needed them— taking out a discretionary loan or suffering no loss—and least agreeable when a distress loan was needed or when everyone was lodging a claim after a hurricane. A reading of the rhetoric expressed in Brazil, Mexico, and Argentina during 1983 and 1984 about the banks and their evil agent, the IMF, provides images almost identical to those expressed in the nineteenth-century frontier United States about the eastern banks.

The U.S. Treasury—never a real friend of the Federal Reserve System, the central bank of the United States since 1913—brought to Bretton Woods a distrust of Keynes's ambitious and advanced notion of a world central bank. But initially, although the idea of a world central bank disappeared, the notion of a credit pool together with symmetric adjustment responsibilities for surplus and deficit countries remained intact. Moved no doubt by the "covetousness and rivalry" cited by Hammond, the American Bankers Association tried to cut down even the restricted powers of the IMF while the U.S. Congress ratified the Bretton Woods Agreements in July 1945, only after long debates that continue to this day to surround congressional decisions on the IMF or the World Bank.

Unfortunately for the long-run viability of the Bretton Woods system, its actual operation never really represented even the limited concepts of shared adjustment contained in its original form. This was bound to be troublesome since, as a system of fixed exchange rates, the Bretton Woods agreements removed the primary automatic mechanism for eliminating payments imbalances among the great and small nations of the world. All that was left was coordination of each country's monetary and fiscal policies with those of its trading partners. And without the scarce currency clause, the agreements placed squarely on deficit countries' often shaky shoulders all the burden of adjusting to the requirements of external balance. The surplus countries faced no presumption of external pressures on their internal economic policies, while when deficit countries ran out of options to borrow they faced the shameful —it seemed in the world of powerful nationalism that emerged unscathed after the horror of World War II—prospect of being forced by the IMF as a condition for further credit to alter their own economic policies.

The scarce currency clause—the means to put adjustment pressure on countries in surplus—was simply never enforced. The reason for this atrophy of a crucial element in the international adjustment process with exchange rates fixed was painfully obvious. Since it authorized collective discrimination against a chronically surplus country, it meant, in the decade of the "dollar shortage" after World War II, that the United States would be called upon to follow expansionist policies or suffer collective discrimination against its exports. The fact was that most countries, at least by the 1950s, *wanted* the United States to run a deficit, so they could earn some precious dollars to rebuild reserves. Between 1946 and 1953 the United States *gave* the rest of the world about $33 billion in gifts and loans—about $13.4 billion as direct grants and loans under the Marshall Plan. No one was about to claim, even if it had been possible since the plan dwarfed any resources the IMF might offer—that the scarcity of the dollar meant that Europe and Japan should dictate U.S. economic policy or discriminate against U.S. exports. While "Superpower America" was dispersing billions, holding most of the world's gold, and still issuing dollars as good as gold, adjustment responsibilities for surplus countries—there really was only one—seemed an absurd notion. It was thoroughly driven from the minds of governments and central banks during the decade after 1945 and replaced with the basically mercantilist notion that surplus countries are "strong" and worthy while deficit countries are "weak" and due for some well-deserved nasty medicine. The perception of the IMF as an agent of the strong, imposing unbearable burdens of adjustment on the weak, now called "conditionality," has been reinforced—somewhat unfairly—by protests from Brazil, Argentina, and other heavy borrowers over conditions attached to new loans since 1982.

Propping Up the Bretton Woods System After 1960

The immediate postwar "dollar shortage" era changed complexion radically by the late 1950s. We have already seen that large U.S. balance-of-payments deficits began to result from an outpouring of dollars to Europe and Japan. Foreign governments, especially the French, with noisy admonitions from de Gaulle about U.S. abuse of the system, began more and more to convert into gold the dollar IOUs that they accumulated from U.S. balance-of-payments deficits. By 1960 dollars held abroad totaled more than the $19 billion U.S. gold hoard. The London gold bubble of October 1960, triggered by that offhand remark presidential hopeful Kennedy let drop, marked the knife-edge watershed from "dollar shortage" to "dollar glut."

The Bretton Woods system and the IMF in particular were totally unprepared to deal with an actual or perceived surfeit of dollars in the world's monetary system. With no pressure on them to adjust, the strong-currency countries either delayed currency upvaluations until they had accumulated huge reserves, as in the case of Germany and Switzerland, or clung to decades-old exchange values until forced to adjust by the dollar gusher that erupted after August 1971, as in the case of Japan. This placed virtually all of the 1960s adjustment burden on emerging weak-currency countries like England, Italy, the United States, and eventually even gold-centric France, and meant that devaluations, when they came, would have to be larger than would have been the case under symmetric adjustment.

The United States was, however, in a special situation. As long as currencies were pegged to the dollar, as specified in the Bretton Woods Agreements, surplus countries were forced to buy the dollars pushed into their hands by U.S. deficits. Theoretically, the dollars could be converted into gold either directly by central banks at the U.S. Treasury or through direct gold purchases, which indirectly drained U.S. gold through the London gold pool. In practice, the Germans, Swiss, Japanese, and French governments were subjected to a combination of cajoling, rewards, and threats by the U.S. to hold the dollars. The system held together for a decade because the surplus countries liked their surpluses and knew that to end dollar accumulation by way of a sharp dollar devaluation would raise the double specter of an end to the fixed-parity Bretton Woods system and an end to their large surpluses—recall the nervous responses to John Connally's November 1971 suggestion for a 15 percent dollar devaluation. These fears kept the ever-complaining Europeans and Japanese inert until the United States itself decided to act in 1971. The negotiations surrounding this uneasy truce were conducted directly between governments with little effective IMF involvement.

The combination of delayed adjustment by surplus countries coupled with an abhorrence of exchange rate changes placed tremendous pressure on the deficit British during the 1960s. The moral overtones of efforts to "save the pound" had their roots in over a century and a half of British sterling-centricity in the world financial system coupled with its ignominious collapse in 1931 when Britain left gold and with it left Dutch and French central banks with heavy losses. The British, with the aid of U.S. and European partners in the system of fixed parity, propped up sterling well past the time that it might have hoped to survive the 1949 exchange rate of $2.80 to the pound. Heavily restrictive policies that were very costly in terms of British jobs and production failed to validate the overvalued sterling exchange rate. The moral tone of the crusade even reached across the Atlantic to the New York Federal Reserve Bank and Charles Coombs, chief officer on the foreign exchange desk.[8] But economics, not morality, ultimately governs the rate of exchange between monies, and to pretend otherwise is only to delay the inevitable, perhaps itself a kind of immoral act. The first major victim of a flawed Bretton Woods system, sterling was devalued by 14.3 percent in November of 1967, some two or three years too late. The 1971 collapse of that system was precipitated by a U.S. president who had to untie the dollar from gold in order to print the IOUs needed to pay for Vietnam and the Great Society. The result, as we have seen, was a worldwide inflation sufficient to validate the 1973–74 oil price surge.

The IMF Transformed to Adjustment Assistance Agency for Developing Countries

The 1960–70 decade of travail for the world's money system revealed how painfully far from reality had fallen Keynes's dream of the IMF as a world central bank. After the 1960 gold bubble flashed the signal that the system was in trouble, there was no direct open involvement of the IMF as a central banker. Instead the finance ministers and central bankers of the ten major industrial countries—Belgium, Britain, Canada, France, Germany, Italy, Japan, the Netherlands, Sweden, and the United States—formed their own club called the Group of Ten—the G-10. The G-10 had indirect ties to the IMF through the General Arrangement to Borrow (GAB), under which G-10 members could, at their own discretion, supplement IMF resources.

The emergence during 1960–62 of the G-10 and GAB effectively relegated the IMF to two decades during which it held power largely over developing countries in search of short-term adjustment assistance and over occasional errant, deficit industrial countries like Britain. Regular American participation at BIS meetings began in December of 1960, with attendance by Charles Coombs and later by higher level officials of the Federal Reserve. The BIS became a useful meeting place for the ad hoc G-10 meetings to arrange measures to shore up the system of fixed exchange rates—a kind of anonymous and secret Swiss clinic for treatment of an important patient, knowledge of whose serious illness is deemed not to be in the public's best interest.

From Bretton Woods to Basel

There were only three key words, articulated by the "Group of 32" (there are many "groups" in international finance)—economists who met at a 1964 conference at the beautiful village of Bellagio overlooking Lake Como in northern Italy, about an hour's drive north of Milan: liquidity, adjustment, and confidence. Liquidity meant central bank reserves adequate to deal with the ever-increasing ebb and flow of foreign exchange markets that had to be offset if exchange rates were to remain fixed. Adjustment meant—effectively, given the structure of the Bretton Woods system—belt tightening by deficit countries with the exception of the United States, which could issue dollar IOUs to pay for its deficits. Confidence meant quite simply confidence in fixed exchange rates and in a fixed dollar price of gold.

As the 1960s wore on and strains on fixed exchange rates built, it was at the low-key and quiet BIS meetings of central bankers where solutions to the triad of liquidity, adjustment, and confidence problems were worked out. G-10 finance ministers—the Treasury Secretary in the United States—were not welcome at the BIS; it is a central bankers' club and central bankers view finance ministers as politicians. Fritz Leutwiler, president of the BIS, has made clear his feelings about politicians: "I have no use for politicians. They lack the judgment of central bankers."[9] Notwithstanding Leutwiler's views, central bankers do talk to their respective finance ministers and heads of treasuries. And their regular low-key BIS sessions gave them the subtle "feel" of things in the world financial community at large that could be passed on to finance ministers especially during times of crisis, like the revaluations of the German mark and the Dutch guilder in March of 1961 or the sterling devaluation in November of 1967.

Besides these adjustment episodes, BIS discussions backstopped the evolution of an enhanced network of central bank liquidity and confidence measures. The London gold pool to keep the price of gold at $35 an ounce in the wake of gold bubble or other jitters was arranged with the background provided to G-10 ministers by BIS discussions. General Arrangements to Borrow and Swaps—credit lines between central banks—were also put together in this way. A steady flow of background consultations at the BIS, combined with G-10 meetings when crises flared, supplied more liquidity when it was needed, shored up confidence when it sagged, and, most important, began to evolve the notion that there were times when even exchange rates had to move. The ad hoc nature of BIS and G-10 arrangements allowed the system some flexibility to evolve with the changing needs of the international financial system. That system was moving in a direction that was to put more and more weight on the "adjustment" leg of the Bellagio triad.

United States economic and financial hegemony was eroding while at the same time the $35-per-ounce price of gold was becoming more and more of a fiction in which it was difficult to place much confidence. In 1934, when FDR set the price of gold at the $35-per-ounce level, the U.S. consumer price index stood at about 40 (with 1967 = 100.) Over three decades and a world war later, in 1955, it reached twice that level, and by 1967 it had risen to two and one-half times its 1934 level.[10] That made gold at an unchanged price a very good buy relative to a loaf of bread, which in 1967 was well over twice its 1934 price. And gold producers found that an ounce of gold, which cost over twice as much to produce in 1967 as it had in 1934, bought a lot less than it had three decades earlier. Little wonder that the supply of newly mined gold virtually dried up. Supply had to come from gold hoards, and ultimately the only seller consistently willing to provide gold at $35 an ounce was the U.S. Treasury. In short, supply was drying up except for that which drained gold out of Fort Knox while demand was rising steadily. Eventually the dollar price of gold would have to rise, and Franz Pick—publisher of *Pick's Currency Yearbook*, which annually foresaw the demise of $35-per-ounce gold—and his "gold bug" cronies knew it.

From on High

The IMF Goes for SDRs

While the system gained momentum in the swing toward adjustment and the BIS and G-10 ad hoc operations were learning, painfully and slowly, to roll with the punches and gaining the experience that would eventually help to produce a new system, the G-10 was operating in conjunction with the IMF on another much more highly publicized front. For its part the IMF set its sights squarely on the goal of enhancing liquidity to shore up the rigid exchange rates and gold price at the heart of the Bretton Woods system. In July of 1964 the G-10 set up a study group on creation of reserve assets. What emerged three years later at the September 1967 Rio de Janiero Joint Meetings of the IMF and World Bank was the Special Drawing Right (SDR), somewhat ineptly dubbed "paper gold."

The SDR was the IMF's baby. Here at last it seemed was a reserve asset for central banks that could be created at the stroke of a pen. If the world's finance ministers and central bankers got together and agreed to accept SDRs in settlement of debts accumulated among themselves from balance-of-payments imbalances, then indeed SDRs were reserves—"as good as gold," it was sometimes hopefully said. The SDR plan was a way to increase the IMF's resources to deal with the periodic currency crisis that in its view arose from irresponsible actions of speculators trying to cash in on a forced devaluation. In principle, if the central banks held virtually unlimited reserves that they could create on their own initiative, speculation against fixed exchange rates would be severely discouraged. The IMF, which was to be the repository and distributor of SDRs, seemed on its way back toward realizing Keynes's vision of it as a true world central bank.

The IMF-as-world-central-bank dream, fondly cherished by some as it may have been—was not widely held and was probably not foremost in the minds of many in the trenches dealing with problems of the international monetary system in 1967. Just two months after the Rio meeting had approved an outline for the SDR facility, the devaluation of sterling shook the foundations of a liquidity-based system without exchange rate adjustments. For some, sterling's break away from an exchange rate that had been defended not only by the British but by all of the G-10 countries—$3 billion of credits by the G-10 and BIS in November of 1964, a $1.4 billion drawing from the IMF in May of 1965, increased Swap drawings from the Federal Reserve and other central banks in September of 1966 and July of 1967—was a sign that economic forces ultimately did determine exchange rates.[11] This group, which tended to include economists with less of a stake in the fixed rate system than their colleagues in the central banks and finance ministries, saw the sterling break as the death knell of the Bretton Woods system. For others at the central banks and treasuries, the sterling debacle was a tragedy that need not have happened if only they had had more reserves with which to quell the greedy appetites of speculators who sought to force exchange rate changes for their own benefit. Considerable animosity grew up between these two groups, with the economists characterizing their official colleagues as myopic and the latter characterizing economists as irresponsibly giving encouragement to speculators with their irrelevant theorizing about floating exchange rates.

In the "real world," as they liked to call it, the central bankers and finance ministers at first won out. Arrangements for SDR allocations went forward, and by March of 1968 the G-10 had—with French reservations, of course—resolved final issues on establishment of the SDR "facility," as it came, antiseptically, to be described. The agreement called for creation of $9.4 billion—they had to be measured, prospectively at least, in terms of dollars, not a good sign for an asset meant to replace the dollar as international reserves—to be "allocated" approximately in thirds on the first days of 1970, 1971, and 1972. The $9.4 billion would add about 14 percent to international reserves of industrial countries, which stood at about $66 billion in 1970. (Initially one SDR equaled one dollar).[12]

Liquidity Was Not the Problem

Meanwhile, signs were appearing that the IMF's thrust toward liquidity was leaving badly unattended some problems of adjustment and confidence. The mark was showing signs of being undervalued—priced too low in terms of other currencies—and the Bundesbank was selling a lot of marks and taking in huge dollops of foreign currency at the bargain-basement price dictated by the official exchange rate. During the strongest surges the Germans took in $2.8 billion in November of 1968 and $4.0 billion in early May of 1969. On May 9, 1969, the exasperated German cabinet rejected revaluation "for eternity."[13] "Eternity" lasted a little over five months—until October 24, to be exact—when the mark was revalued by 9.3 percent after having been allowed to float on September 29. Meanwhile, on August 8—again during that dead, sultry period in the summertime when everyone is off on vacation or wishing he was—the French slipped in a franc devaluation of 11.1 percent.

From on High

The French move was particularly significant as it dramatized the end of the de Gaulle era, which had come officially just months before in the spring when the government of General de Gaulle had toppled after years, succeeded by Georges Pompidou and his able Finance Minister Giscard d'Estaing. De Gaulle had been a powerful exponent of the France-as-major-world-power view, jealous and antagonistic toward the United States' postwar position of political and economic hegemony. In monetary matters the French leader had been a great champion of gold primacy—often converting French dollar holdings into the yellow metal so much dearer to the hearts of Frenchmen than bits of paper embellished with pictures of George Washington. De Gaulle also, like any postwar leader who remembered the shame attached to currency devaluations and Britain's departure from gold in the 1930s, equated devaluation with weakness and national disgrace. Throughout the 1960s France had been losing badly in the economic race with Germany. Despite heavy reserve losses, which signaled a need for a French devaluation in 1968—speculators had turned to the franc as their next victim after sterling—de Gaulle denounced devaluation of the franc as the "worst form of absurdity" on November 13, 1968, in the midst of a European foreign exchange crisis that saw all but the most naive running out of francs and into marks.[14] Even a fiery November 24 nationwide address by de Gaulle in which he steadfastly refused again to devalue the franc, as if the honor of France hung in the balance—generals tend to see an assault as something to be resisted to the end—failed to stem the tide. Lyndon B. Johnson, the leader of another country with a somewhat beleagured currency, strongly supported de Gaulle's refusal to devalue. Nine months later a new French government—suggestive of the likelihood that the need for a franc devaluation was not the only urgent economic problem gainsayed by de Gaulle—quietly nodded to reality and devalued the franc.

The SDR initiative was in some ways as much a tragedy for the world financial system as it was for the IMF. To beleagured finance ministers and central bankers, the late 1960s were a nightmare. Like harried doctors in an emergency room after a terrorist bombing, they only had time to treat symptoms. Injured victims kept crying out for more time to adjust, for more transfusions of reserves. When, under the terms of the Bretton Woods system, the reserve increments came in the form of dollars pouring out of the United States to finance its deficits, they cried out in pain. That was the wrong medicine, thought the men in the emergency room. Led by the French, the loudest cry was for more reserves of gold. Desperately they extemporized to produce paper gold: SDRs. Every patient—every chronic "weak" deficit country like France, Britain, or Italy—came in crying for more reserves, and so the decision was made to synthesize gold. Less notice was given the "strong" surplus countries like Germany, Japan, and Switzerland, countries that, by definition under the terms of the Bretton Woods system, never appeared as emergency room patients. They were to be given something nice into which they could convert their ever-increasing stock of dollars.

The SDR initiative had the IMF on a sweeping run around the liquidity side of a line that it seemed was blocking evolution of a viable world money system when the right call was to run to the opposite, adjustment side. The goal for the IMF was realization of its potential as a world central bank, a supplier of the reserve asset that national central banks will want to hold and use in settlement of "temporary" payments imbalances between themselves. A question, too painful to address in the midst of the currency turmoil at the end of the 1960s, was whether the quantity of reserves on hand might not tend to color governments' perceptions about which imbalances were temporary and which were not and thereby indicative of a need for exchange rate movement. For men who reject revaluation "for eternity" or devaluation as "the worst form of absurdity," either the imbalances had automatically to be prejudged as temporary or the reserves available to shore up the existing exchange rate had to be infinitely large. It was often forgotten that the inflows to the "strong," "well-behaved" surplus countries that perceived no urgent need to adjust unless the inflow of money threatened to flood them with money and cause inflation—that is, to force them to adjust—mirrored outflows from the "weak" and "errant" deficit countries that threatened to exhaust their reserves. German, Swiss, and Japanese surpluses only mirrored French, Italian, British, and U.S. deficits, so that refusal either by surplus or deficit countries to adjust amounted to pressure on reserves of deficit countries.

Too little adjustment created the necessary preconditions for creation of SDRs. The deficit countries wanted more reserves, while the surplus countries, a little less anxiously, wanted a superior reserve asset into which they could convert their "unwanted" dollars. Paper gold seemed the answer. Those with reservations were largely consoled by the thought that at least SDRs held out the promise of avoiding the unimaginable compound calamity of breaking the fixed dollar link to gold and slipping into the utter chaos of floating exchange rates. Many at treasuries and in bank boardrooms around the world in 1970 truly believed that the world economic system would cease to exist without fixed exchange rates. During a 1966 debate with Milton Friedman, Robert Roosa—former Treasury Undersecretary, then a partner at Brown Brothers Harriman—actually denied that a foreign exchange market would exist under floating rates.[15] How could anything so strenuously resisted at such great cost, and strongly advocated by "irresponsible market crazies" who knew nothing of the "real world"—like Milton Friedman and Harry Johnson—be anything but total disaster? It was, of course, necessary to exclude from the "real world" Canada, whose currency had floated for twelve years, from 1950 to 1962, and then had been refloated on June 1, 1970.

The architects of the SDR facility—as they noncommitally called it since from one quarter or another there always existed objections to calling it anything tangible; you always were being told what it was *not*, never what it was—discovered that merely wishing for a gold substitute does not make it materialize. What was really wanted was a reserve asset for central banks. Fearing a loss of control akin to the burgeoning of billions of eurodollars, they did not want SDRs in private hands. But a viable reserve asset had to be something that the central banks wanted to hold. And it had to coexist, at first at least, with other reserves: gold and dollars. Surplus countries were careful to specify limits on the quantity of SDRs that deficit countries could foist upon them. A low, fixed interest rate paid from deficit to surplus countries in SDRs on SDR holdings by designated surplus countries was instituted to sweeten the deal for sometimes-hesitant "strong" currency countries and to remind deficit countries of their obligation to adjust.

The simple truth was that no one really knew what to make of an asset that existed only in the vacuum of central bank vaults and that could be defined only in terms of the national monies—dollars in particular—it was supposed to supplant. Efforts were made to redefine it in terms of currency "baskets," and some long-delayed adjustments were made in interest rates, but the SDR was always a kind of anemic cousin to dollars and gold, which, for all of their faults, were at least tangible and real, in use every day as money and/or assets.

The SDR experiment might have been a fascinating one in synthesizing a money asset simply by common agreement among central banks. Money, after all, is money only because it is generally accepted by fiat in payment of debts. But its value in terms of other things, established through exchange in open commerce for tangibles, is ultimately what makes people willing to hold it, and no one wants to be paid in something he does not want to hold until he in turn pays someone else. Lacking the hard-to-define but very real aura of acceptability attached to widely held and used monies like the dollar and gold, the SDR looked something like the "cheap money" that, under Gresham's Law, drives the "dear money" out of circulation. Like a torn or frayed bill or a Canadian quarter in the United States when the Canadian dollar is worth only 80 cents, it was the first thing one spent, especially for tips or donations where one feels that the recipients—like deficit countries— ought to be grateful for whatever is offered.

Looking Back Instead of Ahead

The events of 1971 and afterward showed that the SDR and with it an IMF committed to fixed parities propped up, if necessary, by an ever-increasing supply of "paper gold" were badly out of step in the inevitable march away from rigidities imposed two and a half decades earlier. Besides being vaguely unacceptable, oriented as it was toward possibly inflationary increases in liquidity, the SDR facility held the possibility of exacerbating the inflationary pressure that was present throughout the 1970s. Amid the rush of more pressing matters, the facility was never extended after the last of the initial set of distributions was made on January 1, 1972. The Committee of Twenty— an enlarged version of the G-10 constituency—negotiated steadily and tortuously from the autumn of 1972 to June of 1974 in a pathetically persistent attempt to revive a modified Bretton Woods system: one in which exchange rates could move but not by very much and only when a set of preordained criteria said they could. By June of 1974 the events in exchange markets and elsewhere had given governments a sufficiently heavy dose of reality to prompt them to put the C-20 out of its misery.

The SDR liquidity thrust and the dream of realizing early hopes for an IMF as a world central bank were sidetracked from the international financial system's center stage after August of 1971. Recall IMF Managing Director Schweitzer's invitation to join John Connally in his Treasury office to view President Nixon's televised "New Economic Program" announcement—almost an afterthought, like inviting neighbors to a party only because they are close by and might be offended by knowing you had left them out. The IMF was not to reemerge on central stage until over eleven years later, when, on November 16, 1982, a group of shocked bankers meeting at the Federal Reserve Bank of New York learned from another Frenchman, Jacques de Larosière, that the IMF was not about to bail them out of Mexico but rather proposed to bail them *in* to the tune of $6.5 billion, payable, if you please, within thirty days, by December 15. Have a pleasant Thanksgiving, gentlemen.

Looking Toward Europe

The IMF had by 1971 become only a symbolic embodiment of some sense of cohesion in the postwar international monetary system. In a world of powerful nation states, first faced with nearly absolute U.S. economic hegemony and later with resurgence toward a significant European role—the Japanese had still not decided to exercise a major role in determining the shape of the world's money system—"supra-national" organizations like the IMF, United Nations, and World Bank turned out really to be "subnational." They were, after all, no more than creations of their member governments whose resources were drawn entirely from those governments, especially the richest among them, which, in turn, expected to exercise control proportional to the size of their contributions. Moreover, for the Europeans as well as for the Third World, the IMF is an American creature, or at least an Anglo-Saxon one, reflecting its origins inside the U.S. and British treasuries. Its treasury origins make it also a political creature, because treasuries after all are charged with providing governments with the financial means to operate and that means accommodation of government borrowing and debt servicing at minimum cost. These goals are often at odds with sound money objectives of central bankers who sometimes must allow interest rates to rise high enough to signal a need to end the government spending in excess of income that so often follows from wars or, since Keynes, from hopes that demand might somehow create its own supply.

For the men—and they are only men—in Europe concerned with the soundness of money, its central bankers, the IMF was too American, too open, and too close to its treasury origins. For them the only institution approaching a supra-national central bank was the Bank for International Settlements that, after all, had been in existence since May 1930, although, as mentioned earlier, it did not hold its first annual general meeting until the inauspicious month of May 1931, just as the failure of Austria's Kredit Anstalt became known. The BIS was the bridge from the old prewar international financial system to the difficult transition underway in the interwar period. Its members may have seen it as taking over responsibility for maintaining cohesion of the world financial system after the failed British-U.S. efforts of Montagu Norman and Benjamin Strong to put the world back on a sterling-based gold standard at the prewar rate of exchange between British pounds and gold.

The twenties had been a U.S. decade that saw Britain's newly ascendant cousin trying to help restore the Old Lady of Threadneedle Street to her prewar position of preeminence. But the 1925 return to the gold standard and the prewar parity of $4.86 to the pound was—as we have seen—not sustainable, even with Benjamin Strong's rapid expansion of U.S. money, which only served to fuel an unsustainable boom in security prices.

The founding of the BIS in May 1930 was coincident with the start of a decade during which what international financial leadership there was came largely from it. After going off gold in September 1931, the British were in no position to lead. FDR specifically eschewed any such role for the United States after the London World Economic Conference in June of 1933. The BIS had by 1935 held its fiftieth session, and more than half of the central bank governors had not missed a single one.

Not until 1944 at Bretton Woods did the British and Americans try to resurrect a system that they conceived and dominated. In the coming postwar milieu the politicians aimed to reassert themselves over the bankers. New Dealers' populist, U.S. antibanker sentiments surfaced at Bretton Woods and an attempt was made—based on a trumped-up charge that the BIS was laundering gold the Nazis had stolen from occupied Europe—to abolish it. The initiative was firmly put aside, but the signal could not be missed. After the war, the United States intended to be in charge, with a new institution that had no ties to the byzantine past of European finance. The BIS, for its part, simply continued its monthly meetings of central banks, not including any participation of the Federal Reserve. (Prevented by congressional suspicions about banks, especially mysterious Swiss ones, the United States Federal Reserve was not among original purchasers of BIS shares. United States participation was confined to a group of commercial banks whose shares were held in trust at the First National City Bank, now Citibank.) The very regularity of BIS meetings built up a reservoir of goodwill and mutual understanding among the central bankers in attendance that could be drawn upon in times of crisis—almost like the cohesion of a small town or village community that grows in part out of regular meetings at churches or social organizations.

The BIS After World War II

With its continuity and coherence, the BIS became a key player in restoration and maintenance of the postwar world financial system. In 1950 it became the technical agent—handling international payments bookkeeping—for the European Payments Union, established in May of that year to begin rebuilding an intra-European payments mechanism. On a worldwide scale, the BIS played its first crucial role by arranging the gold pool and central bank currency swap mechanisms after the October 1960 gold bubble and the March 1961 currency market turmoil surrounding revaluations of the German mark and Dutch guilder. Both the gold bubble and revaluations were signs of dollar weakness, the first against a commodity standard and the second against other currencies—a revaluation of the mark and guilder can, as we have noted already, just as well be seen as a devaluation of the dollar. When postwar U.S. financial hegemony—wrested jealously away from the IMF during its infancy —came into question in the watershed year of 1960, the atrophied facilities of the IMF required that some alternative had to be found. The BIS was the natural place, and since its meetings were held monthly, no nerve-jangling "special" meeting had to be arranged. Neither did the central bankers working on the crisis have to come together as quasi-strangers. With the exception of the U.S. Federal Reserve's representative—appearing regularly only after 1961 —attendees at the Basel Club's monthly meetings were old friends.

As a "happy coincidence" that only comes from thinking ahead over decades, the March 1961 meeting of the BIS started on the evening of Friday March 10 just a few days after the German and Dutch revaluations. After the meeting the assembled central bankers issued a rare statement to the press, stating with remarkably simple forthrightness that they "wished it to be known" that they were cooperating closely in the exchange markets.[16] What this really meant was that they had set in motion arrangements to establish an intricate network of currency swaps—preset agreements between central banks to lend each other foreign exchange in time of need. The idea behind a currency swap is simple. If the Bank of England is running short of German marks, speculators will think twice before betting against sterling if they know that the Bundesbank—the printer of marks—is prepared to lend large amounts (billions of dollars' worth) to the Bank of England.

The informality—no highly trumpeted economic summit meeting among heads of state—simplicity, and secrecy of BIS operations causes them to be almost overlooked, which is precisely what the publicity-shy bankers want. No attempt at further interpretation of the seemingly innocuous BIS statement came until six months later and, even then, not from a widely read source. On page 10 of the September 1961 Bank of England *Quarterly Bulletin,* the aim of the swap arrangements was cryptically described as being that "no party of the so-called 'Basel Agreement' should be forced by speculative movement of funds to deviation from the declared policy."[17] Not to be pinned down even by this single sentence with everything substantive crammed between the line(s), French finance minister Wilfred Baumgartner epitomized the spirit of the BIS at a press conference at the IMF's September 1961 Vienna meeting when he eschewed even the appellation "Basel Agreement" so lightly applied by the Bank of England. *"Il n'y a jamais eu de gentlemen's agreement de Bâle, mais il n'y a á Bâle que des gentlemen."*[18]

Mr. Baumgartner's elegant, arrogant assertion of the European banker's view of the world, though he was himself a mere finance minister, signaled a new phase in the evolution of the postwar world financial system. It was similar in tone to a reply made by Sir Roy Harrod, Keynes's biographer, to a question raised in 1967 by *enfant terrible* Robert Mundell at a University of Chicago seminar after publication of Lytton Strachey's biography had clearly indicated that Keynes was a homosexual—a detail missing from Harrod's reminiscences. Said Sir Roy, "It was there between the lines for those who knew."

"Those who knew" at Vienna heard Mr. Baumgartner thumbing Europe's nose—especially the distinctively prominent Gallic nose of one M. de Gaulle —at the United States, which, it was clear, was going to need the good offices of the BIS to keep the dollar sound. The irony of this in view of U.S. efforts to destroy the BIS in 1944 was noted by its president, who observed with some rancor that "the United States which has wanted to kill the BIS suddenly finds it indispensable."[19] No accident that regular U.S. attendance at BIS meetings began in 1961 with the level of attendance rapidly rising to include Federal Reserve board members and the chairman of the board himself. To this day, however, since U.S. membership on the BIS board would require an unlikely act by an ever-suspicious U.S. Congress, U.S. participation at the monthly meetings is essentially as a guest of its board.

"U.S." and "European" Decades

Ascendancy of the "gentlemen" at Basel in 1961 marked the first decisive postwar influence of an essentially European organization of central bankers, as opposed to U.S. and British treasuries-dominated influence on the working of the world financial system. While the International Monetary Fund had by 1960 emerged as a flawed realization of a world central bank—even for politically oriented finance ministers and treasury secretaries—the BIS, in contrast, showed decisively that it had been from the beginning, was then, and ever would be a bankers' bank. The turmoil of 1960–61 marked a watershed in the decadal ebb and flow of influence over two dimensions: between Europe and the United States and between treasuries, always pressing for easier money and central banks, often trying to resist and finding it easier to do so as a group than individually.

Viewed broadly and retrospectively, the decades since 1950 can, like the interwar decades, be characterized as "U.S." or "European" epochs in international finance. The 1950s together with the postwar 1940s were "U.S." decades with most of the world's monetary gold in Fort Knox and dollars sought all over the world. The 1960 gold bubble put the United States on the defensive in a "European" decade. In August of 1971 President Nixon again seized the initiative for a U.S. "mini-decade" until November of 1978, when a worldwide run out of dollars put the United States back on the defensive. The U.S. inflationary bias, telegraphed to the rest of the world by the Carter administration's advocacy of the "locomotive" approach to world economic recovery and growth by force-feeding demand with easy money, was eventually reversed, tentatively at first, in October 1979 with initiation of the Fed's experiment targeting money supply instead of interest rates, and then decisively, in the spring of 1981, when the Treasury, representing the newly elected Reagan administration, played the unusual role, for that institution, of pressing for strict adherence to money growth targets.

A particularly interesting aspect of this ebb and flow of European/U.S. initiative is the underlying trend, until 1980 at least, of rising European concern—much of it originating from monthly gatherings of hard-currency men in Basel—with reining in a U.S. tendency toward expansionary money policy. During the 1960s it took the form of a—later justified—concern that too many dollars worldwide jeopardized the dollar's link to gold. After August 1971, with the exception of a brief period after the first oil crisis, European concern with expansionary monetary policy—unfettered by a tie to gold—again justifiably grew. It culminated with formation of the European Monetary System (EMS) during 1978–79, where German and French fears that the dollar was finished as viable international money were sufficiently strong for them to overcome their traditional animosity and to push for a European-organized and controlled monetary system as a hedge against the dollar's total collapse.

Primacy of concern about inflation—deep concern borne out of painful interwar experience—has until recently in the postwar period reflected two basic ingredients: "European" and "central bankers" (as opposed to finance ministers or treasury secretaries). The BIS, formed in the interwar period and growing in influence ever since despite early U.S. foot-dragging, quintessentially captures these two ingredients.

The Contemporary BIS

Viewed from the vantage point of the United States in the 1980s, the BIS is still a strange bird. In an era of strong governments and distrust of banks, it is a private banking company—its shares trade on the Paris exchange—yet it can accept deposits only from central banks. It is virtually immune from government interference under guarantees of an international treaty signed in The Hague, Holland, in 1930. The BIS continued to operate in Basel throughout most of World War II, even though its membership included, among others, belligerents Germany, Italy, Britain, and France, a fact that re-emphasizes that it is an organization of central bankers, not governments, with the former existing on their own, apart from governments. While elections mean that governments come and go and wars may move boundaries and change allegiances, eventually they will end and commerce will revive. That revival will require a return to familiar monetary arrangements whose existence the BIS stands ready to preserve.

The BIS jealously guards its low profile. Only reluctantly did it in May 1977 move out of its old, unmarked headquarters in a deserted Basel hotel to a circular, eighteen-story "Tower of Basel," which looks like a nuclear power plant with windows. The current president—until the end of 1984—Dr. Fritz Leutwiler, bemoans the new headquarters, which have become something of a tourist attraction on the otherwise-bland Basel landscape: "If it had been up to me, it never would have been built."[20] Conspicuous as the new building is, even a visit to the lobby requires a pass. And a pass requires specific business at the bank, which you are unlikely to have unless you are highly placed at a central bank or part of the BIS staff—a staff whose ranks extend from exotic researchers to cooks and chauffeurs, all of whom serve its directors.

Almost like an old man set in his ways, the BIS does not alter itself to changing times. Monthly meetings ten times a year on the second weekend of the month—August and October are skipped—bring central bankers into Basel for Friday-night-to-Sunday-evening sessions from all over the world. Though air connections to Basel are not the best—London or Paris would be much simpler—the BIS remains in Basel, where fewer financial reporters are likely to be lurking. To move would change its character, and that is the last thing this bastion of conservative, hard money wants.

Although the BIS clings to its accustomed ways of operating, the developments in the world financial arena that it undertakes to be the first to monitor are the newest and most exotic. New and pressing problems always find their way to the desks of BIS governors before they go anyplace else. During the 1960s, when the eurodollar and eurocurrency markets were growing by leaps and bounds, causing great worry for the national authorities whose control these markets had escaped, it was the BIS that devised a massive cooperative effort among the major banks in financial capitals to keep track of their total "euro" deposits. This meant monitoring operations like Morgan's dollar business out of London and Paris and Barclays dollar business out of New York, always attempting to adjust downward the gross totals for eurocurrencies to eliminate interbank deposits and obtain an accurate measure of the eurocurrency systems' net addition to international liquidity. Such deposits do not add to total deposits since they just amount to bank A's deposit in bank B, which is largely offset by bank B's deposit in bank A. The BIS has continued to be the sole source of comprehensive data on gross and net eurocurrency assets and liabilities of the world banking system, data it reports annually along with its views on the world economy in its excellent *Annual Report,* published every June between gold-colored covers.

BIS and Herstatt: Lender of Last Resort?

The BIS's steady monitoring of eurocurrency markets made it a natural locus for examination of extended lender-of-last-resort facilities across international borders in the wake of the June 26, 1974, closure of I.D. Herstatt in Frankfurt, a closure for which the Deutsche Bundesbank (Germany's central bank) did not assume responsibility—that is, it did not make good any losses of banks dealing with Herstatt.

The Herstatt failure involved a clear case of fraudulent operation of a not-very-large bank operation. Bankhaus Herstatt had been speculating heavily in foreign exchange—the same bugaboo that was to bring down the Franklin National Bank of New York a few months later—and losing. The losses were concealed with the aid of modern technology: computer records were altered. Among its responsibilities, Herstatt was a clearing agent for many of the world's large banks, which meant that large sums of money owned by the likes of Morgan's, Bank of America, Chase, and others were always passing through it. When the Bundesbank ordered Herstatt to close down at 3:30 P.M., it was 10:30 in New York, shortly after the start of the U.S. banking day. When U.S. banks routinely called for collectibles due from Herstatt, it was too late. Losses totaled close to $500 million. The Herstatt collapse sent shock waves reverberating throughout the world financial system. The amount involved was, however, only about a third of 1982 monthly debt payments by Brazil alone.

Herstatt's collapse was of relatively small scale, but it suggested huge potential problems regarding unclear lender-of-last-resort responsibilities and it came at a time when nerves were frayed over recent oil price increases, as yet unclear recycling responsibilities, exchange rate volatility, and the last stages of the agonizing U.S. process of sloughing off a president in whom the nation had lost faith.

The lender-of-last-resort question is complicated enough for banks and central banks within their own borders, but it becomes a nightmare with myriad possibilities for crossed signals where the question of banks' foreign branches and subsidiaries is involved. The dangers are akin to the situation that arises when a towering fly ball drops toward the infield on a windy day, first toward the pitcher, then the second baseman, then the shortstop. Occasionally the ball simply drops to the ground, leaving all three players with an exasperated "I thought you had it" look on their faces. At least, for the disgruntled ballplayers, the problem ends there. If Chase's Frankfurt branch was to go belly up while the Federal Reserve and Bundesbank bickered over who ought to step in to deal with a serious liquidity or solvency problem, the problem could snowball. In view of the Bundesbank's less-than-forthcoming attitude regarding coverage of liabilities of one of its own fraudulently operated banks—to be imitated eight years later to the tune of $1.6 billion by the Bank of Italy's abandonment of Banco Ambrosiano's foreign subsidiaries—the implications for its attitude toward foreign branches and subsidiaries of non-German banks operating in Germany were not comforting to central bankers at large. The problem was really an offshoot of the eurocurrency markets, which represented the sum total of operations worldwide by foreign branches and subsidiaries of banks operating outside the borders and therefore away from direct supervision by their respective headquarter country central banks. In 1974 the markets' gross size was $395 billion, with net deposits estimated at $220 billion.

Confronted with this problem and with memories of Kredit Anstalt and 1931 in their minds, the governors of the Group of Ten central banks and Switzerland scurried as quickly as possible to their own lender of last resort, of sorts, the BIS. Herstatt collapsed on June 26, just two weeks before the July meeting at which the problem was discussed. With the August break before them, there was time to allow the BIS staff, together with research staffs in each central bank, to prepare background studies for a fuller discussion. After their September 1974 meeting the governors issued a communique:

The governors had an exchange of views on the problem of lender of last resort in the Euromarket. They recognized that it would not be practical to lay down in advance detailed rules and procedures for the provision of temporary liquidity. But they were satisfied that means are available for that purpose and will be used if and when necessary.[21]

In effect, the governors were saying "it may not look as though the central banks have got our act together yet, but we have committed plenty of funds to manage any problem that might come up." This gave the impression that the BIS was prepared to operate as a lender of last resort in the euromarket at a time when responsibilities of national central banks for their foreign branches and subsidiaries were not well established. Isolated problems like Herstatt, on the order of hundreds of millions of dollars, were what was contemplated, not systemic problems running into hundreds of billions, such as questionable loans to developing countries. In 1974 loans to developing countries were not generally viewed as problems and totaled less than $50 billion.

Formation of a Banking, Regulation and Supervisory Practices committee followed issuance of the BIS September 1974 communique. By December 1975, the Cooke Committee, as it had come to be known, issued the high-sounding "Basel Concordat," endorsed by the governors at their December 1975 meeting. The concordat assigned supervision of liquidity (adequate funds to pay current bills) to "host" authorities: all banks operating in England were subject to Bank of England supervision on liquidity matters. Responsibility for the more serious matter of solvency (assets adequate to cover all liabilities) was split. Foreign branches were to be the responsibility of parent authorities, meaning that Citibank's London branch was monitored by the Federal Reserve regarding matters of solvency, while foreign subsidiaries' solvency was monitored by the host authorities—the Bank of England would monitor solvency of Citibank's foreign subsidiary in London.[22]

The categories designated by the concordat are not watertight, and its guidelines—there were no "teeth" to force adoption—were not fully implemented. Efforts to clarify responsibilities further were intensified in 1983 following heightened concern over situations arising from ownership of LDC loans by foreign branches and subsidiaries. The problem of enforcing guidelines remains—essentially having been left to the discretion of the governor's group meeting monthly at Basel.

Lender of Last Resort Since Herstatt

Since the 1974 Herstatt scare the eurocurrency markets have continued to grow, reaching a gross (net) size of $2 trillion ($1+ trillion) by 1982. The unresolved problem of clarifying central bank responsibilities with respect to foreign branches and subsidiaries, many of whose assets and liabilities make up eurocurrency loans and deposits, remains pressing in view of the fact that these banks have participated, probably fairly extensively, in lending to LDCs. The danger of an isolated problem arising at a foreign branch or subsidiary is, however, low since trouble with a country borrower such as Mexico or Brazil triggers alarm bells at all the major central banks due to the heavy involvement of their own banks operating inside their borders. Still, eurocurrency operations could intensify problems in a crisis where default or its imminence triggers a heavy need for liquidity to reassure depositors. Central banks might be so pressed in dealing with their home banks that eurocurrency banks might be left to seek help from host central banks, which, in turn, might be preoccupied with their own home banks. Were they so ignored, the eurocurrency banks would not simply drop quietly between the cracks since they have correspondent ties with home banks. Impairment of eurocurrency banks' ability to honor their commitments would very quickly exacerbate liquidity problems of home banks everywhere.

To anticipate such problems, the Cooke Committee endorsed the principle of supervision on a consolidated accounting basis whereby a bank's capital adequacy and risk exposure is evaluated, including positions of all of its foreign branches, subsidiaries, and affiliates. Such evaluations were to be semiannual, but in the light of rapid developments regarding many of the heavily indebted developing countries, a move to quarterly monitoring with as little lag as possible between quarter-end and date of evaluation seems advisable. Provision of comprehensive, timely information is expensive and has presented a serious problem for banks seeking adequate evaluation criteria of their own operations as well as for regulators seeking to monitor overall exposure. The BIS and the 1983 successor to the Cooke Committee are grappling with these problems. As the central banks' banker, the BIS is uniquely qualified to direct coordinated efforts to monitor overall exposure.

The "Inner Club"

The tremendous power of the BIS to deal with crises and ongoing problems in the international financial system comes not from any great wealth it possesses, although its assets are considerable and have grown rapidly during the

From on High

1970s. Their total is not known, being expressed as "billions of gold francs" with no exchange rate into a more familiar unit of account. The real power of the BIS lies with its ability to focus and mobilize the assets of the central banks—all of the largest and most powerful ones—represented regularly at its meetings. Numerous odd divisions of responsibility at the BIS preserve the history of involvement in its varied operations. The board excludes the United States, Japan, and Canada, whose economies account for some 40 percent of world economic activity and dwarf the economies of the European governors on the board. The gold pool still meets monthly, as do the G-10 and the European Community. But the real power is wielded by the undefined "inner club" of the BIS, composed of the heads of the world's largest central banks who ultimately must deal with the world debt crisis that began in 1982.

A senior member of the inner club is Karl Otto Pohl of Germany's Deutsche Bundesbank. He commutes to Basel meetings from Bonn in his Mercedes limousine rather than flying in a government plane—perhaps to demonstrate his bank's cherished independence from the German government, perhaps to savor the pleasant ride along 325 miles of autobahn south from Bonn along the Rhine to Basel. For a busy man like Mr. Pohl, such a drive can provide precious undisturbed time in which to prepare for the upcoming Basel weekend. Along the way one can stop at some of the world's finest restaurants, perhaps the Auberge de L'Ill at Illhaeusern whose *noisettes de chevreuil aux champignons des bois* has earned it three Michelin stars.

The British envoy to the Basel inner club is Gordon Richardson, retired governor (head) of the Bank of England. Lord Richardson is the only member of the inner club who is not currently head of a central bank (or on its board, as in the case of the Fed's Henry Wallich). It may be that the BIS does not deem Britain's Thatcher government sympathetic enough to its aims of supranational cooperation that in times of crisis transcends national interest narrowly defined—the British have, among the major powers, been least supportive of rescue packages for Mexico, Brazil, and other problem borrowers since 1982. Alternatively, the British government itself may prefer to have Lord Richardson, whose longstanding involvement with the BIS reaches back to early sterling and dollar rescue days of the 1960s, as an envoy whose stature within the BIS is high while still being subject to qualification at home. In effect, Lord Richardson's unofficial standing together with the official but less powerful role played by his successor as governor, Robin Leigh-Pemberton, gives Britain considerable flexibility in defining the Thatcher government's somewhat ambivalent relationship to the BIS.

The regular U.S. representative to the BIS inner circle is Federal Reserve Board member Henry Wallich, although the Board's chairman, Paul Volcker, becomes the prime spokesman whenever he chooses to attend a Basel weekend.

Mr. Wallich's rather special status of being an inner circle member without being a head (or former head) of his central bank comes from his longtime involvement as a liaison between the United States and Europe on international financial matters. Before being appointed to the Federal Reserve Board in 1974, Mr. Wallich was a Yale Professor of Economics and manager of the U.S. Treasury's academic consultants groups on international financial problems. His profile was raised considerably by a stint as a regular contributor to *Newsweek*'s weekly column on economics. Berlin born, his contacts with other European-born economists transplanted in the United States—including Yale colleagues Robert Triffin and Willy Fellner, Fritz Machlup of Princeton, and Gottfried Haberler of Harvard—were always close, and over the years his regular attendance at BIS meetings—where German is the language of intense discussions—has built up his standing. President Carter's appointee to the chairmanship of the Federal Reserve Board before Paul Volcker, G. William Miller—not a man of sufficient stature in financial circles to merit the privilege—was never admitted to the BIS inner circle. In that situation, or during the doubtful period in 1983 before an ambivalent President Reagan reappointed Volcker to a second term as Fed chairman, the ongoing presence of European-American Wallich was reassuring. Continuity and predictability are valued highly at the BIS, an agency whose standards are an allegory for the necessary preconditions of an orderly monetary system.

Although their institutions are somewhat less independent of sitting governments than the Bundebank or even the somewhat beleagured Federal Reserve, Lamberto Dini of the Bank of Italy and Haruo Mayekawa of the Bank of Japan are also inner circle members. Italy's presence in the inner circle owes much to the influence of Guido Carli, a past head of the Bank of Italy who was a widely acknowledged senior statesman of international finance during the 1960s and 1970s. Carli provided decisive leadership—devising, among other things, a two-tier gold market in 1968 that kept official gold at $35 per ounce while allowing private trading at a market-determined price.

Mayekawa seems, from the outside, a one-dimensional presence—"Japan is so big they really have to be there"*—in the very European inner circle of the BIS. But in a way he is the quintessential European central banker, utterly circumspect and placing a high value on continuity and predictability, which enable consensus decision making. As a much older society than any in Europe, Japan has refined further the delicate requirements for maintenance of useful trappings on highly civilized cultures of which money, especially paper money, is one—being no more than an idea possessed of value unrelated to its physical content and determined solely by the care taken by those in

*This phrase, frequently heard in Western circles, has been used to dismiss the Japanese while acknowledging their economic power.

charge of its issue. Beyond his Japanese qualities, which make him a "natural" for the BIS inner circle, Mayekawa is a highly competent and visionary leader of the Bank of Japan who is very skillfully and cautiously taking necessary steps to increase its independence from the all-powerful Japanese Ministry of Finance. Part of his strategy is to discreetly raise the international profile of the bank as the world's most powerful trading nation moves toward a role in the international financial system commensurate with its role in the world economy. Recent liberalization, in December 1980, of rules controlling Japanese international borrowing and lending, together with the heavy involvement of the traditionally highly leveraged big Japanese banks in loans to Mexico, Brazil, and other developing countries, makes this internationalization of Japan's outlook and its central bank highly desirable and appropriate. In recognition of this and of Mayekawa's high personal standing with his fellow heads of central banks, he is included in the inner circle.

Paul Volcker, as we have said, occasionally chooses to attend a Basel weekend. When he does attend he is afforded a warm welcome and a place of honor at the inner council's table. Such treatment is not due to his position at the head of the world's most powerful central bank—sometimes even called the world's *de facto* central bank. Volcker's eminence is due rather to his longstanding association with the key men in the international financial community, which reaches back to the late 1960s when he was Treasury Undersecretary for Monetary Affairs in the Nixon administration. In a U.S. administration not very sympathetic to—or sophisticated about—international problems—as it was to demonstrate clearly with Nixon's New Economic Program in August of 1971—Volcker alone seemed to understand European concerns about gold, the dollar, and exchange rates. During the 1950s he had been an economist at the New York Federal Reserve Bank—later to return as its president—and had also worked a stint as an economist at the Chase Manhattan Bank. He knew the banker's point of view and he knew what was on the minds of the Europeans late in the 1960s when the Bretton Woods system was coming apart.

Beyond his uniquely cosmopolitan view of world financial matters, Volcker has the bearing of a banker. He listens. At seminars and meetings with economists from inside and outside the Treasury, Undersecretary Volcker absorbed everything that was said—sometimes heatedly—about the pros and cons of exchange rate flexibility, the roles of gold and the dollar in the system, and what should be done about the ever-widening U.S. balance-of-payments deficits. He sat, puffing on his cheap cigar, and spoke very little during or after the sessions, usually leaving everyone a little uneasy. Many thought they were there to win an argument, to show Volcker that they were logically correct and thereby convince him of the optimal path for U.S. policy to follow on international issues. But Volcker was and is an eclecticist. He wanted from those sessions a review of options on the issues, which he could then compare with political constraints imposed on solutions that the economists had not included in their models. He also could report to the nervous Europeans the state of American economists' thinking on the issues and give an assessment of how influential (not very) that thinking might be.

Volcker is the quintessential central bank head. He has over the years held all the right positions: New York Fed economist, Chase economist, Undersecretary of the Treasury, and president of the New York Federal Reserve Bank, which serves as the primary international operative for the Federal Reserve System and for the U.S. Treasury. Having been inside the U.S. government (as opposed to its central bank) in his high-level Treasury position, his perspective as a prognosticator for the European central bankers as to what the Americans (the U.S. government) will do next is invaluable. As U.S. administrations from Nixon to Ford to Carter to Reagan come and go with exits and entrances of a whole new cast of characters at Treasury, State, and Commerce, Volcker can tell his Basel friends something about what baffling new round of U.S. policy initiatives to expect.

The crowning glory to all this, of course, is Volcker's remarkable record since he took over as chairman of the Federal Reserve Board in August 1979. A desperate Jimmy Carter turned reluctantly to Volcker—fearing, correctly as it turned out, his reelection-threatening independence—as literally the only man who by merely taking the job could begin a U.S. antiinflation program with a presidential demonstration that "this time" he was really serious. After only two months on the job Volcker, having recognized that a clear demonstration of different Fed procedures was required to provide convincing evidence of an ongoing, substantive initiative to control inflation, announced on Saturday, October 6, a new set of Federal Reserve operating procedures.[23] The new procedures, which have been mentioned already and will be discussed fully in chapter 10, were widely billed as a "monetarist experiment" whereby the Fed concentrated only on hitting preannounced money growth targets, paying no attention to interest rates, which previously had been primary targets of Fed policy.

This simplistic interpretation was wrong, as anyone who knew Volcker ought to have known. What he was really doing—brilliantly, as it turned out —was gaining the Fed some elbow room by untying it from exclusive targeting of interest rates, announcing targeting of money growth while operationally retaining the ability to target *both* money growth and interest rates. Volcker knew that slowing down inflation would mean a money-crunch period during which interest rates would skyrocket. If, under new operating procedures, the Fed could say it was not targeting interest rates, the inevitable congressional uproar that always accompanies serious efforts at inflation control might be deflected long enough for the program to work.

The rocky road traveled by the Federal Reserve from the fall of 1979 to the onset of the debt crisis and thereafter is a part of our story soon to come. For now it suffices to say that, at least in the view of his Basel colleagues, who have surely not agreed with his specific policies at every step of the way, Paul Volcker has acquitted himself as a man of great vision, principle, and courage. At the core of this respect is an admiration for his ability to maintain Federal Reserve independence from an ever-meddling U.S. government, from the always-antagonistic Congress to the White House and Treasury. That trio blows hot and cold, supporting inflation control when it is politically expedient and, when it is not, shamelessly leaving Volcker alone to defend himself against criticism of policies designed to lower inflation. That goal was, after 1980, noisily espoused by an American president and his Treasury Secretary along with the usual coterie of congressmen, all of whom appear to have had little idea of what it takes to achieve it. Above all else, Volcker's noble forebearance and independence in the face of government pressure have won him a place of honor at the inner council of the BIS that will, if he desires, outlast his chairmanship of the Federal Reserve Board.

Head of the Inner Circle

If Paul Volcker is the quintessential head of a national central bank struggling for its independence from government interference, Fritz Leutwiler, president of both the Swiss National Bank and the BIS, is the quintessential world central bank head. The Swiss National Bank, Switzerland's central bank, is—like the Bank of England was from 1694 to 1948—privately owned. Moreover it is, unlike any other central bank, virtually autonomous, required to answer no questions of Parliament or finance ministers. This independence likely accounts for the stability of the Swiss franc as a storehouse of purchasing power during the otherwise chaotic 1970s. Every wealthy investor would like to have some money in Switzerland, but there is not much room in the tiny economy, even with its huge bulging financial muscles. The Swiss know that the best form of reserve is a widespread, unsatisfied desire to put more money into the hands of its circumspect bankers.

As head of the BIS inner circle and its president, the tough, athletic-looking

Leutwiler is described in a *Harper's* magazine article as "arguably the world's most powerful banker."[24] His worldwide power is not derived from any immensity of BIS assets or power of the Swiss National Bank. Rather it derives from the simple fact that he presides every month over meetings of the heads of the world's most powerful central banks, meetings that shape the rules of the world money game. BIS meetings strongly influence bank governors' views regarding where they think the world money system should be moving on crucial issues like exchange rates, flexibility, coordination of economic policies, LDC debts, and a host of other issues that define the arena within which the players in the system must operate.

The BIS meetings have played a crucial role in shaping responses—both immediately and over the longer run—to the problem of coping with inability of LDC borrowers to service their debts. When Jesus Silva-Herzog, on August 13, 1982, confronted the shocked U.S. with the fact that Mexico did not have the funds necessary to avoid default, Paul Volcker's first call was to Fritz Leutwiler, who—inevitably, in the European dead month of August—was hiding from the swarms of tourists in the little Swiss mountain village of Grison. What was needed was an immediate $1.85 billion to get Mexico out of borrowing problems in the interbank market where its ability to keep rolling over a huge volume of overnight loans had come to an end.

Leutwiler arranged a temporary bridging loan for $1.85 billion on the telephone in less than forty-eight hours; largely it came from the central banks of the inner club. Half came from the United States, $600 million from the Treasury's exchange stabilization fund—which somewhat ironically was created in 1934 out of profits from the U.S. devaluation of the dollar against gold from $22 per ounce to $35 per ounce—and $325 million from the Federal Reserve's accounts. The Mexican bridge loan was a classic demonstration of the power of Leutwiler and the BIS. The financial resources of the BIS per se were not involved. The keys were, first, the reservoir of cooperation and understanding that comes out of face-to-face monthly meetings and, second, the absolute discretionary power of Leutwiler. Unlike Jacques de Larosière, whose IMF has great resources and power in its own right, Leutwiler has no board of directors that he must bully or cajole into approving his actions. Even if de Larosière handles his board well, he cannot act until notice of a board meeting(s) is duly given and, sometimes lengthy and often multiple meetings have been held. There was no time for any of this in the Mexican crisis. Leutwiler was the only man in the world who could have received Volcker's call late on a Friday in August and had $1.85 billion ready to tide Mexico over by the time banks opened on Monday morning. Silva-Herzog was playing a dangerous game. By letting Mexico's payments problem reach a crisis stage, he had lit a fuse that only one man could put out in time. Lucky for him that

From on High

Fritz Leutwiler was not off on a weekend hike on Friday, August 13, 1982.

The revelations that followed the Mexican crisis left the BIS inner circle with virtually no choice but to continue as managers of a burgeoning worldwide debt crisis. In 1983 the BIS extended a $400 million loan to Brazil. Results of it were not as satisfactory as those achieved with the Mexican loan, which was repaid on August 31, 1983. By July of 1983 the Brazilians were due to repay the $400 million and failed to do so even though Leutwiler had announced in advance that the BIS would not extend the payment deadline. By August, when the Brazilians missed another deadline—the July deadline was extended in spite of Leutwiler's warning—the exasperation of a powerful yet frustrated man showed through. In an interview with the Zurich newspaper *Tages-Anzeiger* on August 20 he spoke with incredible candor, betraying a Swiss banker's impatience with borrowers not living up to their obligations: "Things just cannot continue as they have been. These [debt] problems will never be solved . . . with money and more money. . . . To say that these [debtor] countries should not be treated with toughness is absolutely grotesque."[25]

Mr. Leutwiler's lament was the cry of the central bankers who made up the inner club at the BIS: if the only way out of the debt crisis was believed to be "more and more money" and the inflation that follows, then all the progress made since 1980 in reversing the trend toward ever-spiraling world inflation would be lost. That would be too great a price to pay for enabling the LDCs to honor contracts drawn up in anticipation of just such an inflationary spiral. It would only mean that such contracts would be drawn up again and again, followed again and again by pressure for more inflation to keep the music going until finally, as it had half a century earlier, the music would stop and debts of essentially bankrupt countries would be deemed worthless. A repetition of the familiar path to financial collapse would be shattering to the close-knit circle of the world's most powerful central bankers. Having seen the whole thing happen once before from their vantage point in the little Swiss city nestled against France just south of the beautiful countryside of Alsace-Lorraine, they would find severely shaken their passionate belief that for civilized men who can learn from experience to modify their behavior there is nothing inevitable about recurrent financial collapse. The philosopher Santayana has said, "Those who do not learn from the past are doomed to repeat it." The men who gather every month at Basel have learned from the past but, powerful as they are, they may still not be powerful enough to prevent its repetition by others who have not learned as well.

PART V

A Spring Uncoils

Chapter 10

Wishful Thinking
and Bad Luck

THE 1979 annual meetings of the IMF and World Bank were unique in two ways. They were held for the first time in Belgrade, about forty miles from the Iron Curtain, which separates eastern Yugoslavia from Rumania. Holding the meetings in a country whose economy is planned but whose market elements and politics are unaligned with either the communist or capitalist bloc of nations seemed appropriate for organizations whose operations span the full spectrum of both the economic and political. The Yugoslavs, flushed with over a decade of prosperity, had built to house the meetings the Sava Center, named for the river that crosses their nation from westward-looking Ljubljana close to the Italian border to eastward-looking Belgrade.

The Belgrade meetings were also unique by virtue of the unusual behavior of one of their most distinguished and highly visible attendees. Paul Volcker, for less than two months head of the world's most powerful central bank, left Belgrade to return to Washington at the start of the formal meetings, not staying even to hear hapless Treasury Secretary William Miller "praise the 'courage and intelligence' of President Carter's economic policies in his formal address to the meetings."[1]

Widespread Dissatisfaction with U.S. Policies

Less than a week before fleeing the Belgrade meetings on Wednesday, October 3, Volcker and Miller had left Washington for the conference. Their first stop was in Germany, where they planned to put forward the usual pleas for patience with U.S. efforts to engineer a "soft landing" whereby monetary policy gradually reigns in inflation without precipitating a recession. Volcker undoubtedly expected that he still had some goodwill he could draw on with Bundesbank president Otmar Emminger. But the Germans, along with their European allies and the Arabs, who as heavy accumulators of dollars over the half decade since 1974 now had considerable clout, were having no more of U.S. gradualism and soft landings. The U.S. delegation left Bonn, and the flak intensified as soon as they reached the Sava Center.

Tension about the disastrous economic policies of the United States had been building all summer in the capitals of Europe as well as in Tokyo and Riyadh. Even U.S. embassy personnel, besieged at their posts by dismayed representatives of host governments, were themselves deeply troubled by Washington's "business as usual" attitude, which belied its total failure to perceive the panic that was building about the future of the dollar. Carter somehow thought that he had made his concession to the foreigners by appointing Volcker as Federal Reserve Board chairman in August, against his own wishes, some believed.

But the Arabs and the Europeans, who had all along been baffled and unimpressed by the first truly regional American President—is his name really Jimmy?—were by the summer of 1979 totally out of patience with his economic policies. Evidence was mounting that the U.S. commitment to the "dollar rescue package," which Carter himself had announced, was waning. Money growth, which had slowed to about a 5 percent annual rate in the first quarter of 1979, had reaccelerated sharply. From April to July it shot up again to a 12 percent annual rate. Through all of 1979 U.S. inflation ran at well over a 12 percent annual rate on its way to a December 1979 level that was an incredible 13.3 percent above the level of the previous December; hardly the sort of performance that would reassure foreign holders of U.S. dollars who had been ready to dump them over a year earlier, in October of 1978, just before the ill-fated "dollar rescue package."[2]

United States policy was badly misdirected on two fronts. On the energy front, Carter's absurd program of price ceilings kept U.S. gasoline prices at about one-half the typical world price to consumers. On the monetary front, the Federal Reserve had been pegging the Federal Funds rate at which the banks lend to each other below 10.25 percent during much of the second quarter of 1979, and this had resulted in the burst of rapid money growth. The Fed's pegging exercise resulted in a 10.2 percent interest rate on Treasury bills and a 9.33 percent interest rate on ten-year Treasury bonds during September of 1979. With inflation running at over 12 percent, "real"—inflation-adjusted—interest rates were running at minus 1–2 percent, and Americans were borrowing like crazy to buy second houses, diamonds, Chinese ceramics, land, antiques, and anything else they thought would appreciate along with or ahead of the dizzy spiral of prices.

For the Europeans, it was hard enough to endure the disgusting vision of Americans gobbling up underpriced oil along with everything else they could lay their hands on with money borrowed at negative interest rates—with the interest all tax-deductible to boot—while their somnolent president talked about gradually bringing down inflation without precipitating a recession. But what they found intolerable was the tremendous spillage of dollars outside of the United States that resulted from the government-subsidized orgy of U.S. consumption engineered by the Carter team. In May of 1971, when Europe had first begun to choke on U.S. dollars, the total of those dollars outside the United States stood at about $55 billion, having grown at about 10 percent a year during the decade since the gold bubble had first signaled European dollar indigestion. In the eight years between August of 1971 and August of 1979, freed from the discipline of gold, U.S. monetary profligacy had spewed dollars into OPEC coffers and central bank vaults at a rate of over 20 percent a year, more than twice the rate that had worried everyone so much under the old dollar-exchange standard in the 1960s. That system had at least demonstrated a semblance of concern about maintenance of the dollar price of gold.

The U.S. locomotive, which pushed up U.S. demand without increasing the supply of anything but paper money, dumped an additional $100 billion dollars onto the rest of the world between 1976 and 1979. By the time Volcker was virtually booed out of Belgrade, foreigners were holding more than a quarter of a trillion dollars—$267 billion, to be exact—and, as far as they could see, were earning a negative return of 2 or 3 percent on the largest part of it—the part that had been accumulated over the years when the U.S. commitment to price stability and positive returns on Treasury bills and bonds had been believed. A negative return of 3 percent on $267 billion in dollar assets meant annual losses of $8 billion on dollars held to make it easier for Americans to buy subsidized gasoline and live in houses twice the size of European houses, thanks to extra low, tax-deductible interest payments on their mortgages.

Of course, all that easy U.S. money meant also that lots of Americans could borrow cheaply on the huge increase in the value of their houses to splurge on German Mercedes or BMWs or perhaps to buy, as a second or third car, a Toyota or Datsun. On a more basic level, to be accumulating the billions of U.S. dollars, the Europeans and the Arabs had to be buying them—no one was actually forcing them to do so and the dollars were not just appearing out of nowhere at the Bundesbank. Heavy dollar purchases by the Bundesbank, the Bank of Japan, and other central banks were the only thing preventing the dollar from depreciating even further—from March of 1977 the dollar had depreciated against the mark by a third, against the yen of an oilless Japan by 26 percent, and against the once-swooning British pound by 29 percent.[3]

Viewed from the vantage point of exporters in those countries, a United States on a consumption binge with underpriced foreign exchange and negative real interest rates was just the ticket—as long as your central bank kept shoring up the dollars in which you got paid and as long as you could keep raising U.S. dollar prices of your wares sold in the United States fast enough to make up for more dollar depreciation that was always just around the corner. Mercedes and BMW put through so many price increases for their U.S. lines—many of them in excess even of the dollar's depreciation against the mark, "anticipating further depreciation of the dollar," as they liked to say—that keeping up on current prices of the wonders of German engineering got to be like following prices of corn futures during a rainy summer. You knew they would be higher, but they never ceased to amaze you with just how high they went. (As if by another engineering miracle, the prices somehow managed to free themselves of the pull of gravity. They never came back down when by 1981–82 the dollar's weakness became the dollar's strength, so that still they levitate as if by magic at levels some 40 percent above prices of the same cars in Europe, thanks largely to the nontariff barriers provided by the U.S. Department of Transportation and Environmental Protection Agency.)

But while exporters in Germany, Japan, and elsewhere were very happy with the U.S. dollar in 1979, governors of central banks were not. The flood of dollars was putting heavy inflationary pressure on their economies. Even finance ministers who, as politicians, loved those whopping export sales, knew that they were beginning to see too much of a good thing. In inflation-paranoic Germany anything approaching double-digit U.S. inflation figures is political suicide.

But the governors like the Bundesbank's Otmar Emminger felt trapped. If they just stopped buying dollars the rate of exchange between dollars and marks or yen would plunge. Even the possibility of this had already during 1978 and 1979 caused a lot of OPEC money to shift out of dollars and into marks. A surge of tens of billions of dollars into Germany would make their inflation problem intolerable, even given the limited ability of central banks to offset (sterilize) the impact of money inflows or outflows on the home money supply. In 1979 the German money supply roughly equivalent to U.S. M-1 (currency and checking deposits) stood at around 230 billion marks ($100 billion.) With $250 billion sloshing around outside of the United States, if a mere 10 percent of it tried to run into Germany for cover from a plummeting dollar, the German money supply would explode by 25 percent virtually overnight. The dollar "overhang" problem, as it had come to be called, was way too big to solve just by letting exchange rates adjust.

In effect, the problem stemmed from three years of not having allowed exchange rates to move enough. The Germans and the Japanese, from their point of view, had been buying dollars to smooth out what they expected, with some encouragement from the Carter administration, to be a temporary weakness. As it turned out, the weakness persisted and worsened, so that dollar holdings built up abroad like billions of tons of water behind the floodgates of a dam during a prolonged rainy spell. To just throw open the floodgates would result in disaster, but something had to be done since leakage over to the top of the dam was beginning to eat away at the very foundations that held it in place. The floodgates had already been opened by as much as anyone dared and though the dollar was depreciating, it was not falling fast enough. Something else had to be done to lower the pressure.

The Substitution Account

As they headed for Belgrade late in September of 1979, the word on the minds of all the central bankers and finance ministers was "substitution"—give those holders of a quarter of a trillion dollars something else into which their dollar overhang could be converted. For the Germans and Japanese, the idea was attractive because they did not want the dollars rushing into marks or yen thereby complicating the task of domestic money management and causing intolerable inflation. Besides, Emminger had his hands full keeping the fledgling European Monetary System together. Ominous signs of increasing "diversification" out of dollars into marks were already appearing. The Bundesbank estimated in 1979 that the share of marks in the world's total of foreign exchange reserves had risen from 6.9 percent at the end of 1977 to 10 percent in September of 1979.[4] OPEC finance ministers, growing restive with mounting dollar accumulations that paid negative real interest rates, were moving more and more new receipts—oil buyers were and are invoiced in dollars—into marks. If they decided to shift massive past dollar accumulations into marks, the fears of Emminger and others trying to control German inflation would be more than realized. Both the nervous Europeans and Japanese and the restive oil ministers thought they had been given some encouragement on the substitution idea by U.S. Treasury Undersecretary Anthony Solomon, who had made some favorable noises on August 27 in a speech before the Alpbach European Forum in Austria: "A gradual reduction in the dollar's relative role would appear consistent with underlying developments in the world economy, and that prospect, if it materializes, does not cause difficulty for the United States."[5] Solomon's statement—or understatement—must have caused some snickers. It was like saying to a bunch of hungry construction workers living on Spam at an isolated worksite for six months that they could, if they liked, switch to roast beef.

The substitution account, repository of the idea to switch out of dollars into "something else" but not marks, yen, or gold, was as flawed a notion as the idea to "solve" the world debt crisis by allowing the banks holding billions in LDC loans to exchange them for claims on the IMF or some other collective sugar-daddy. The holders of billions of dollars in 1979, like the holders of billions in LDC loans three years later, were stuck with assets worth considerably less—in "real" terms, the goods and services they could buy—than they had paid for them. Getting out from under that unpleasant fact meant finding someone—or, failing that, something—willing to pay up full list price for merchandise selling on the open market for 80 to 90 cents on the dollar.

Architects of the substitution account hit on the idea of converting "excess" dollar balances outside of the United States into SDRs. The trouble was that nobody knew how big excess dollar holdings were or—far more crucial—how just renaming dollar balances SDRs was going to restore the real value they had lost, especially in view of the fact that SDRs in their 1979 form carried interest rates set fairly rigidly at only 80 percent of the weighted average of short-term interest rates on the five major countries. (Part of the European's proposal was to get the United States to give exchange rate guarantees plus interest subsidies on SDRs to make them more attractive. The United States was not inclined to go that far.) Besides, the SDR was essentially composed of the currencies of those five major countries—by the end of 1980 the sixteen-currency basket was abandoned for a more realistic five-currency basket that gave the U.S. dollar a weight of 42 percent; the deutschemark, 19 percent; the French franc, Japanese yen, and pound sterling, 13 percent each. This odd weighting scheme, which gave the Japanese yen a weight equal to the French franc or British pound and less than the German mark while Japan's GNP exceeded that of Germany and Britain combined, reflected the odd combination of European presumption and Japanese diffidence that has characterized world monetary arrangements since World War II and that will soon have to change.

No matter how you looked at it, the substitution account was a muted way to get restive dollar holders into something else like German marks since it proposed SDRs, themselves nearly half dollars while only a fifth marks. But the widespread support for the substitution account and the deliberations to effect it, underway seriously since early 1978, were a means to confront the United States with well-organized opposition to its easy-money, heavy-consumption policies that—as everyone overlooked at the time—were also sustaining the LDC lending boom. After Paul Volcker—who knew something was up but was unprepared for the barrage at Bonn and Belgrade—had been in the very different and bracing atmosphere outside of Washington, D.C. for a few days, he knew that substitution account or not, the United States had better come up with something that pretty dramatically flashed to its allies a determination to put its monetary house in order. For starters, he simply bolted Belgrade and headed for Washington three days before the meetings ended and—as we have noted—even before the hapless U.S. Treasury secretary, Bill Miller, got to give his address of reassurance to the assembled delegates. Volcker was literally and symbolically putting some distance between himself and the errant Treasury.

Dramatic Announcement of Fed Policy Change

While the markets fretted on Friday, October 5, that Volcker had resigned—as perhaps he indeed had threatened to do to get the attention of those who had not been abroad lately—a bold new proposal for the conduct of U.S. monetary policy was being assembled at a feverish pace. The idea—to emphasize control of the money supply over control of interest rates—was nothing new. It had been getting lip service from the Fed for a decade and had been around as an idea since David Hume had argued against the mercantilists for a gold standard over two centuries earlier. But to put it into operation in a few days with the creaking and not-always-responsive Fed open market desk apparatus was quite another thing. But on Saturday, October 6, Paul Volcker went on the air to take the first step.

Volcker's address was brief and decisive, effectively quelling any doubts about his intention to stay on as chairman of the Federal Reserve Board. He sounded determined and very much in control, and unlike the announcement of the long-since-aborted dollar rescue package of November 1, 1978, this manifesto was made appropriately by the Fed chief and not by a president eyeing reelection possibilities only a year away. The discount rate at which the Fed lends to its member banks was raised by a full percentage point, to 12 percent, and banks were again—as they had been in the 1960s—required to hold reserves against liabilities to foreign branches. The latter move closed a loophole whereby U.S. banks got around the Fed's stinginess in supplying reserves by borrowing heavily—and profitably, since unlike domestic borrowing no nonearning reserves were required—from their foreign branches that were tapped into the huge eurodollar reservoir.

The centerpiece of Volcker's new program—and it was his program, having been rammed through his Board of Governors, three of whom had dissented just two weeks earlier from a halting, half-percentage-point rise in the discount rate—was its emphasis on controlling growth of bank reserves, the raw material out of which money is manufactured, instead of the Federal Funds rate. That rate signals the day-to-day degree of stringency in the market where banks borrow and lend to adjust their reserve positions. The Fed had been setting vague money growth targets for years, but they never had been taken very seriously because no workable mechanism was in place to implement them and, more pragmatically, the Fed's past actions had consistently demonstrated that it did not take such targets at all seriously. The latter fact of Fed life made it very difficult to convince the households and businesses of the world that the guardian of a sound dollar would suddenly begin again—as it had over a decade earlier—to take its responsibility seriously. The Fed had been too concerned with stabilizing interest rates to control the money supply.

Early Stages of the Fed's New Program

Two things can be said of the early stages of what eventually turned out to be the Fed's serious inflation-control initiative. First, the program was not executed cleanly as a gradual but steady reduction in the growth of money, and second, no one—including the Fed itself—really knew what it would mean for money and credit markets.

The extreme volatility of both interest rates and money growth that had followed inception of the October 1979 experiment, not to mention the eerie, unfamiliar "feel" of financial markets that the new operating procedures had created, made it difficult to shake the impression that anything could happen —an impression abhorred by market traders with heavy exposure. Market participants found it difficult to forget that on Monday, October 8, 1979, when their leaders gathered in New York before opening up the first day of business under the new procedure, Peter Sternlight, the Fed's man in charge of the domestic money desk, had shocked the group by saying in effect that he "didn't know any more about the new procedures than they did" and that the Fed was going to be "feeling its way" under the new procedures.

"The journey of a thousand miles begins with but a single step." The Fed set out to convince the world that it was indeed serious about controlling inflation. On the way it encountered many obstacles: an abortive credit controls program during the second quarter of 1980 imposed by a panicky Carter administration that set both interest rates and money growth gyrating wildly; stubbornly high interest rates that the initiative was designed eventually to lower; strong resistance from the Europeans and Japanese—who had at first pushed hard for the United States to control its inflation—to the ice-water shock of a real and persistent slowing of money growth begun in April of 1981; a prolonged and unusually severe U.S. recession in 1981 and 1982 coupled with the highest interest rates in U.S. history during 1981; and, finally, the world debt crisis, which actually began to unfold during 1981 but was not publicly acknowledged until August of 1982.

In a very real sense the world debt crisis of 1982 had begun over a decade earlier when President Nixon cut the dollar tie to gold, thereby clearing the way for the Fed to try, once too often, to print its way out of a recession. Then came the oil supply shock with its overnight creation of lenders with billions to lend and borrowers needing to borrow billions and with the banks, hungry for growth and ready to sell loans, right in the middle. Add to this the fruitless U.S. attempt of 1977–79 to inflate its way out of recession, and you have a U.S.-led world on a credit-buying binge reminiscent of the late 1920s.

False Starts on Inflation Control: Prelude to Disaster for LDCs

Between October of 1979 and March of 1981 Volcker and the Federal Reserve were forging a chain of synthetic gold with which to shackle down the runaway inflationary psychology of borrow-buy-boom and then borrow-buy-boom some more, with only the suckers who saved their money left behind. From mid-1981 to mid-1982, as the Fed pulled harder on the chain, it squeezed hardest against those who had bet the new synthetic link—the will to control inflation instead of a mechanical link to gold—would break. But the chain held, and among the hardest squeezed were the banks and their heavy-borrowing clients in Mexico City, Rio, Buenos Aires, Caracas, and Manila.

The 1970s were characterized in a retrospective article published in December 1979 by the *Economist* as "the OPEC decade." In a prospective discussion of international banking in that issue, the *Economist* looked ahead and made some hard points:

There is no single banking authority in charge of the international markets. The markets are undisciplined (witness the scramble when Iranian assets were frozen). Their rules are untested and often unclear. Long before Iran, central bankers had been voicing concern about—and later discussing collectively—the unbridled growth of international banking. Not all agreed control was needed and nobody knew whether it could be effective. On the agenda for the 1980s?[6]

In this atmosphere of doubt and a search for leadership in controlling inflation and settling credit markets, Carter and his shaky team of economists panicked and imposed the already-mentioned credit controls in March of 1980. The immediate result was to create a vicious whipsaw effect in the financial markets. Credit controls ration money. Once the ration is used up no more is available at any price. Therefore, when credit controls are announced, the initial response by those with commitments to supply credit—the loan sellers—and those with nondiscretionary, absolute needs for credit is a desperate rush to satisfy their requirements before it is all gone. This response causes a sharp rise in rates. Immediately after the imposition of controls, the Federal Funds rate soared from 14 to over 18 percent. Such a movement of over four percentage points in less than a month by a rate that for the previous ten years had ranged typically between 5 and 8 percent, topping 10 percent only in 1974 and 1979, was devastating to the sense of equilibrium in world financial markets. Once the initial panic for funds died away in April, the depressing effect of controls on economic activity dried up the demand for funds and rates plunged, dropping below 9 percent by June. At the same time the money stock plunged, pulled down by a sharp drop in demand. Credit controls were ended by June of 1980, but not before serious damage had been done to the credibility of the inflation control program, and the banks and their LDC borrowers thought they had seen the Fed wink, as if to say "go ahead."

After plunging to a negative value during the spring of 1980, money growth burst upward again, climbing at an annual rate of over 11 percent between May 1980 and April 1981. With it, interest rates shot up again, reaching 21 percent by January of 1981.[7] By winter of 1981 the markets were very doubtful whether the new administration and the Fed could, even if both wanted to, do much to restore control over money and reduce inflation.

The point of detailing the profoundly uninspiring efforts to establish a credible U.S. monetary policy that plagued the late 1970s and early 1980s is to create an understanding of why the remarkable success achieved by the Fed after April of 1981 in lowering and stabilizing money growth came as a massive shock to the world economy. From the early days of its operation during the winter of 1981, the Reagan administration loudly announced its intention to, among other things, control inflation. For those who might have had difficulty hearing, including Paul Volcker himself, the twanging voice of Beryl Sprinkel,

Wishful Thinking and Bad Luck

the new Treasury Undersecretary, carried the message loudly and clearly. And the Fed delivered, lowering money growth to 4.6 percent, well within its target range, between April of 1981 and July of 1982.

But a quick restoration of confidence in the Fed, together with a widespread conviction that inflation was really going to be controlled, was too much to expect in light of the previous three years experience, not to mention the much longer experience since Lyndon Johnson's 1967 decision to go for both guns and butter at once marked the beginning of a tedious litany about controlling inflation and the even more tedious litany and harmful failure to do so. The "conservative" investor who "played it safe" and bought thirty-year U.S. government bonds yielding 5 percent in 1967 earned a *negative* rate of return of 2.5 percent from date of purchase to 1981. At that rate, a dollar of purchasing power in 1967, rather than having been preserved or enhanced, had eroded to about 70 cents by 1981. And this was the experience with the most conservative investment widely represented to investors as the safest way to store purchasing power.

On the other side of this saver's debacle was the positive experience of borrowers able to repay loans at 70 cents on the dollar and purchasers of homes, antiques, and other real goods, whose value was bid up sharply by those fleeing from financial to real assets.

Against the background of persistently accelerating inflation worldwide and the extreme volatility of the 1970s and early 1980s, it was little wonder that very few decisions taken during the year after March of 1981 reflected a conviction that inflation would slow down, even given the serious pronouncements of the new administration. General Motors readily granted its employees 9 percent wage increases over a three-year contract, as did numerous other companies. Boeing rushed to bring out a new generation of airplanes that were cost-effective in a strong economy with jet fuel at or above $1.50 per gallon; it subsequently fell well below that. The best corporate borrowers continued to be willing to pay over 16 percent to borrow money, and short-term rates surged back toward 20 percent during the summer of 1981. And, most ominous of all, the developing countries increased their long-term debts to banks by over $43 billion and paid interest rates of over 14 percent, meaning that they expected both continued inflation worldwide and strengthening of their own export prices and volumes.

Just as the U.S. Federal Reserve was struggling to forge its new, synthetic chain of inflation control, the results of its past neglect were again emerging in the oil market. During the unsettled period between the summer of 1979 and the end of 1980, the oil princes overlaid the uncertainty and confusion surrounding the Fed's sometimes-volatile "monetary experiment" with another doubling in the price of oil.

By the end of 1981, aided by the sharp curtailment of Iranian supply and a final, counterproductive spurt of money growth between mid-1980 and the first quarter of 1981 in the United States, oil prices had more than doubled over 1979 levels. From levels of $2.9 billion and $39.2 billion during 1978, the combined current account surplus of oil exporters and the combined current account deficit of non-oil developing countries rose in 1980 to $116.4 billion and $86.2 billion. By 1981, as in 1975, the ratio had been reversed, with a combined $68.6 billion surplus for oil-exporting developing countries compared with a $99 billion combined deficit for oil-importing developing countries.[8]

The stage was set for a repeat of the explosion of inflation that followed the events of 1974–75, and more contracts were written in inflation-jaded money, commodity, and labor markets that reflected widespread anticipation of just such an outcome. We have seen that 1981 saw the largest increase ever, $43.3 billion in debt to banks of non-oil developing countries on top of a $40 billion increase in 1980. Interest rates were high and floating on this newly contracted debt, reflecting the widespread anticipation that interest costs would be offset by higher export prices and the banks' expectation that floating rates on LDC borrowing would protect them from risks of even higher inflation. Gross interest payments of non-oil developing countries nearly doubled from $28 billion in 1979 to $55 billion in 1981, while over the same period debt service as a percent of exports rose from 21.9 percent to 22.9 percent. This figure rose sharply in 1982, to 28 percent.[9]

Early Signs of Trouble for LDCs

Part of the reason for the absence during 1981 of a sharp rise in debt service relative to exports was the exclusion of short-term borrowing (maturity of less than one year) from published statistics. Such loans were thought to be self-liquidating and virtually riskless. Though data are scarce, it is clear that short-term borrowing rose sharply during 1981 and 1982 while banks started to hedge on their lending to LDCs. According to estimates by Morgan Guaranty Trust Company, short-term borrowing by developing countries totaled over $130 billion by the end of 1982.[10] This development, together with the sharp run-up in debts on terms that anticipated renewed high inflation during the 1980s coupled with the Fed's sharp pull on its newly forged chain of inflation control in the United States during the year following the spring of

1981, set the stage for a major crisis in the world economy during 1982.

More ominously, the world economy began to slow down, especially for non-oil developing countries whose terms of trade slipped by 4.3 percent in 1980 and 2.2 percent in 1981, while 1981 output growth dropped to 2.5 percent from 4.4 percent a year earlier.[11] Growth of output for industrial countries, the major markets for LDC exports, slowed to 1.1 percent, from 3.6 percent in 1979 and 1.3 percent in 1980. It had been expected to accelerate. The volume of world trade was static in 1981 after rising by 6.5 percent in 1979 and by 2.0 percent in 1980. The combined current account deficit of non-oil developing countries reached a massive gap of $99 billion in 1981, just as oil producers' surpluses, while still huge, were dropping. At the same time the Fed's new chain was beginning to squeeze extra liquidity out of the banks, making it harder and harder for the world economy to breathe.

By the end of 1981, some developing countries began to run short of cash needed to meet interest payments on their long-term debt. Those interest payments had reached $55 billion a year while, at the same time, export prices and sales were dropping. Countries like Mexico, Brazil, Argentina, and Venezuela began borrowing short term, for periods of less than a year, to make these long-term interest payments, while the banks pretended not to notice either this or the billions of dollars being rolled over daily in the interbank loan market. Both the loan junkies and their suppliers diligently looked ahead to economic recovery in 1982, at which time their dependence on short-term debt was to end. Some financially more sophisticated countries like Brazil and Mexico began rolling over billions of overnight loans in the New York market, taking care to spread commitments around so that the immense volume of their borrowing would not be too obvious. Overall, borrowers thought that the short-term loans were all right since they were financing a "temporary" cash flow problem. Lenders—a little nervously, but still eagerly—assented, reasoning that, after all, short-term loans, especially to sovereign governments, were not very risky since you could reevaluate them every six or twelve months. It was also true that everyone was being careful not to add up all the short-term lending, so that it amounted to a kind of off-the-record accommodation to get borrowers through a "temporary" problem without "unnecessarily" alarming the financial markets in a way that, borrowers reasoned, might unfairly push up rates.

The $130 billion or so in short-term debt that accumulated largely during 1981 and 1982 was dangerous—like a leaky gasoline pump at a station that is careless about enforcing no smoking rules. Typically, short-term debts are used to finance trade. An exporter may borrow on funds soon to be received in payment for goods already shipped, while an importer may borrow to prepay shipments in exchange for a good price or availability and pay off the loan once goods are sold to domestic distributors. The key point is that such loans are self-liquidating and involve only the rearrangement through time of payments and receipts that arise in the normal course of doing business. True, such short-term loans can be and frequently are rolled over, but only on the occasion of another future payment to be received by an exporter or another advance payment for goods that will become property of an importer.

As a result of their typically self-liquidating nature at least before 1981, short-term debts were not monitored closely. Regulations concerning concentration of short-term bank lending were, and still are, less constraining than those applied to long-term lending. Short-term debt was not generally included in external debt calculations for developing countries. It was therefore very tempting during 1981 and early 1982 for banks "just this once" to lend short to countries that were having "temporary" problems meeting interest payments on their long-term debt. Such lending also tended to slip through cracks in regulation. It was viewed as temporary in nature since economic recovery was widely forecast. Since overall figures were not kept, banks that did not want competitors to know specifics about their relationships with country borrowers did not officially tell each other what was going on, although anyone who had wanted to look could readily have figured it out.

The Regional Banks Join in on Short-Term Lending to LDCs

There was another potentially very dangerous aspect of the run-up in short-term lending during 1981–82. Many of the loans were put together by the large money-center banks with heavy participation of smaller, regional banks in the United States. The regional banks were, understandably in view of their normal specialization in domestic lending, not very well informed on economic conditions in the developing countries to which they started lending. Had they known that funds were essentially being used to service long-term debts they might have been hesitant, but perhaps not even this knowledge would have bothered them. The rates to be earned were very attractive, the loans were to governments for only six months, the economy was expected to recover, and downside risk seemed virtually nil since, they reasoned, they could just pull out at the end of six months if the situation deteriorated. In the middle were the big money-center banks. In the light of their immense long-term commitments to developing countries, they could not expect simply to walk away from the short-term loans that, in the absence of a strong economic recovery, made up most of what the developing countries had coming in to pay long-term debt-service charges.

By 1982 smaller, regional U.S. banks came to account for about 20 percent of all U.S. bank loans to developing countries. According to the Federal Reserve Board, by June of 1982 U.S. banks ranked tenth to twenty-fifth in size had exposure to Argentina, Brazil, and Mexico alone that totaled 81 percent of their net worth.[12] As we have already seen, the same figure for the top nine U.S. banks was higher, standing at 113 percent. We have also seen that exposure was highest of all at the aggressive large U.S. banks. The *American Banker* estimated that by May of 1983 exposure of the six largest banks to Mexico, Brazil, Venezuela, Argentina, and Chile had reached more than 200 percent of their net worth. The Bank of America had nearly 55 percent of its net worth at risk in Mexico alone.[13]

Morgan Guaranty Trust Company suggests, consistently with normal practice, that short-term debt ought to average about three months' worth of imports.[14] By the end of 1982, out of a total $87.4 billion in the short-term debt of developing countries, about two-thirds of the total was in excess of the figure suggested by the three months' imports rule of thumb.

With a tripling of normal, permissible short-term lending to developing countries, the biggest U.S. banks had built a shaky inverted pyramid on a narrow foundation of short-term lending. They were heavily committed over the long term. When the LDCs, suffocating in the dust of world recession, were unable to service their debt, the big banks lent short with more of their own funds heavily augmented by funds from smaller, regional banks that had no significant long-term commitment to developing countries. If conditions were not improved when the time came to roll over the new short-term bridging loans, the big banks would find themselves having to replace every dollar pulled back by the regionals plus even more if the cash flow situation deteriorated even further. The narrow tip of the pyramid was to have been tempered hard by a dramatic economic recovery in 1982. But it never came. Instead of being tempered hard, the tip began to crumble.

Economic Slowdown Brings Collapse

By the spring of 1982 word was beginning to reach the Federal Reserve Board through its own contacts as well as through presidents and boards of directors of its regional banks that the U.S. economy was in serious trouble. At the same time bank regulators were beginning to detect some alarming prospects for performance on the close to $100 billion in loans of U.S. banks to developing countries. The two problems were, of course, closely related. Recovery of the U.S. economy during the first half of 1982 had been a key assumption in continued lending by the banks to developing countries. But instead of recovering, real output of the U.S. economy contracted at a 5.1 percent annual rate in the first quarter of 1982. By the end of 1982, U.S. economic output had fallen to a level 3 percent below the level achieved during the third quarter of 1981. The difference between this outcome and what had been forecast in 1981 was closer to 6 percent, or about $186 billion worth of U.S. GNP.[15]

The failure of the U.S. economy to recover in 1982 resulted from an almost-always-fatal combination of wishful thinking and bad luck. The wishful thinking arose with regard to the attitude of the Reagan administration about the impact on real interest rates and in turn on economic activity of sharply higher projections for federal fiscal deficits. The bad luck, encountered by the Federal Reserve, was the unprecedented behavior of the demand for money during 1982, which compounded the upward pressure being placed on interest rates by the outlook for higher deficits.

During August of 1981 the U.S. Congress passed ERTA, the Economic Recovery Tax Act of 1981. It sharply reduced tax rates on the basis of the then-fashionable supply-side assumption that the resulting burst of economic activity would more than compensate for revenue losses due to the lower tax rates. The resolute determination of most analysts to do nothing during August combined with the considerable bulk of the volume describing ERTA details to produce a period of temporary calm after the storm that had preceded passage of the bill. In fact, interest rates, which had peaked in June, began to drift downward so that by Thanksgiving of 1981 bondholders must have expressed a little more ardently than they had a year before their gratitude to the Almighty for blessings bestowed.

In December of 1981 the belief of those same bondholders in the existence of a truly merciful God must have been severely tested. The *Wall Street Journal* disclosed a new set of official projections totaling well over $400 billion for federal budget deficits over the next three fiscal years, 1983 to 1985.[16] Interest rates promptly began rising again to levels that, when adjusted for lower actual and projected rates of inflation, meant a higher-than-ever real cost of borrowing. In January of 1981 high-grade corporate bonds yielded about 13 percent, which, assuming an outlook for inflation at around 10 to 11 percent, set real borrowing costs at 2 to 3 percent. By January of 1982 those same corporate bonds were yielding over 15 percent, with an inflation outlook of 6 to 7 percent, raising real borrowing costs to an excruciating 8 to 9 percent.

The big borrowers in Mexico City, Rio, and elsewhere who had been looking forward to and fervently counting on a reduction of real borrowing costs back to around 2 or 3 percent were devastated by this outcome. Projections of the lower rates very likely had seemed prudent in view of the 0.85 percent that real borrowing costs had averaged from 1973 to 1980. The collapse in the purchasing power of their exports during the latter part of 1981 following a collapse of commodity prices together with a sharp drop in the volume of their exports had almost required that those who dared to look ahead be optimists, trusting in faith alone. The dollar's strength further eroded the ability of the LDCs to service debts, especially those—including almost all of the non-oil exporters —whose exports were not largely denominated in dollars. At the same time that all of these factors were combining to reduce the supply of available dollars, the dollars required to service debts, instead of falling in line with hopes, rose sharply. For the short-term borrowing that many LCDs were undertaking to service longer-term debts, real costs reached over 16 percent during the first half of 1982.

Had the hoped-for recovery of 1982 materialized, these problems might have been alleviated, though by no means eliminated. But with long-term real interest rates over 8 or 9 percent for even the best borrowers and short-term real rates over 10 percent, recovery was impossible. At this critical juncture the Federal Reserve ran into some terrible luck on monetary policy.

The Velocity Collapse: Bad Luck at a Bad Time

During the fourth quarter of 1981 money growth had accelerated sharply to an annual rate of over 14 percent. Although the money growth rate over a year earlier was only about 7 percent, its acceleration at the end of 1981 and into early 1982 had pushed the money supply about $10 billion above target levels. In an effort to correct this overshoot, the Fed reduced money growth to an annual rate of only 3 percent from January to June of 1982.[17]

The intensity of a money crunch and attendant credit crunch can be gauged roughly by the difference between the growth rates of money supply and money demand. Usually GNP growth rates slightly overestimate the growth rates of money demand on the assumption that households and businesses on average have tended gradually to lower the ratio of money holdings to income over time as means and incentives to economize on money balances have increased. But sometimes, usually for relatively brief periods of time, the demand for money moves erratically, producing a sharp rise or fall in the desired ratio of money to income.

During the first quarter of 1982, instead of drifting downward at its usual rate of 3 percent relative to GNP, money demand relative to GNP shot up at an incredible annual rate of 12 percent. This was due partly to the inflation slowdown and the sharp increase in economic uncertainty surrounding the long-delayed recovery and partly to plain and simple random behavior displayed by the households and businesses that stubbornly refuse to obey strictly the laws of economics. Combined with a minus 1 percent annual GNP growth rate, the shock to money demand relative to GNP pushed up overall money demand at an annual rate of 11 percent. Even more dramatic is the difference of 15 percent between the minus 4 percent figure that would be predicted based on typical behavior and the actual 11 percent figure that materialized.

But the money supply rose at an annual rate of only 1 percent during the first quarter of 1982. This figure may have resulted from the Fed's best guess as to where to aim money supply growth to maintain stable conditions in the credit markets. It likely reasoned that money demand does often rise faster than usual during late recessions, especially given a sharp drop in inflationary expectation. One percent growth seemed a reasonable compromise figure that would avoid squeezing credit markets further while at the same time serving to nudge money growth back into its targeted range.

The 10 percent difference between the growth rates of money supply and demand would have persisted without the very sharp rise in real interest rates that materialized to close the gap by pushing down growth of money demand. In other words, if the Fed wants to keep conditions in the credit markets on an even keel, it needs to keep money supply growth rates approximately equal to money demand growth rates. Accomplishing this requires correct predictions of money demand growth rates, since the Fed can only control money supply growth with a lag. During periods when money demand grows erratically, such predictions are often wrong and credit market conditions change, as they did during the winter and spring of 1982.

The surprise growth of money demand—or collapse of velocity growth—during the early part of 1982 caused money and credit markets to be tighter than the Fed desired. By the time feedback from the economy reached Fed policy makers in May and June, a good deal of damage had already been done. Although the experience was very painful for the U.S. economy, it was devastating for the world economy and especially for the LDCs, which were facing a sharp increase in borrowing requirements at the very time when their capacity to service debts was most impaired. Such economic irony is the stuff of which depressions are made.

Wishful Thinking and Bad Luck

Is "Fine-Tuning" Possible?

Viewed retrospectively, the events of 1981 and 1982 raised serious questions about the validity of doctrines that suggest that precise control, or "fine-tuning," of the path of economic activity is possible. The Federal Reserve ended up with a reduction in inflation that, while welcome, was considerably more rapid than it had intended, largely due to unintentionally tight money during the first half of 1982. The result was to move monetary policy to a restrictive posture at the very time when a neutral posture was both appropriate and intended. Changes in fiscal policy and the large deficit projections that resulted had placed an impossible burden on monetary policy as the Fed tried to control inflation without precipitating a sharp recession. In the end, the hedged combination of tight money with lowered taxes and no letup in overall government spending pushed up interest rates to levels that squeezed all but the strongest borrowers out of the marketplace.

The instability of money demand and the problems that result from it exemplifies a problem which has underlain the post–World War II policy revolution. Keynesian doctrine suggested that control of aggregate demand would enable economists to reduce the intensity of business cycles. Such doctrine, however, presupposed a precise knowledge of key economic relationships such as the demand function for money. The truth is that while statistical measurement of these relationships is possible, a good part of their behavior is subject to random or operationally unpredictable change that makes precise adjustments of monetary or fiscal policy chancy at best. You don't try to disarm a time bomb with mittens on. The results can be disastrous.

The Debt Crisis Breaks Everywhere

As the U.S. economy struggled through winter and spring of 1982, reverberations were felt intensely in the rest of the world. Exports of LDCs stagnated and fell, while their borrowing costs rose and the maturity of their debts shrank, exposing them to more required payments or renegotiation of short-term loans at the time when their ability to repay appeared at its worst. Almost half of the debts to banks of Mexico, Argentina, and Brazil were due in less than one year.

At the same time there occurred a series of disquieting shocks reminiscent of events preceding the Great Depression of the 1930s. On Monday, May 17, Drysdale Government Securities, an aggressive U.S. dealer, failed in New York. The big loser was the Chase Manhatten Bank, which had lent money to Drysdale and ended up out $285 million. A month later in London it was reported by the *Economist* that "Mr. Robert Calvis, the chairman of Italy's largest private bank, Banco Ambrosiano, was found hanging with stones in his pockets from London's Blackfriars Bridge."[18] Subsequent investigation revealed a missing $1.6 billion among a tangle of Ambrosiano's foreign subsidiaries, particularly in Luxembourg. The Bank of Italy's refusal to accept responsibility for offshore activities of an Italian bank sent funds fleeing out of Luxembourg and into London and New York. The Bundesbank of West Germany had to step in to maintain adequate liquidity in the West German banks concentrated in Luxembourg.

The disquieting possibility that there might be nowhere to hide from a financial crisis arose on July 5, when, back in the United States, the Comptroller of the Currency closed and liquidated the Penn Square Bank of Oklahoma City. While it was a small bank with assets of about $500 million, losses on this bank and other bad investments eventually resulted in the failure of Seattle's First National Bank and its absorption by the Bank of America in July of 1983.

While all of these disquieting symptoms of problems with the world's financial system were emerging, banks were beginning to curtail their loans to the LDCs. Loans to corporations were far more in favor than loans to developing countries. International banks, which had been extending between $12 and $13 billion in new medium-term loans to developing countries, cut back to a rate of $7.95 billion during the quarter starting on July 1, 1982. Even with a slowdown in borrowing, the developing countries' average debt-service ratio (ratio of debt servicing costs to exports) continued to increase sharply due to the even sharper contraction in exports. The rise from about 50 percent in 1980 to over 70 percent in 1982 was accentuated by a sharp rise in short-term borrowing and by higher risk premiums that banks charged borrowers as their ability to repay became questionable. Spreads over the London interbank rate (LIBOR) rose on average by more than one-half percent from 1980 to 1982.[19] This seemingly small increase represents $3 billion a year on a total debt of $600 billion.

These developments preceded the onset in mid-August 1982 of what came to be called the world debt crisis. On Friday, August 13, Mexico closed its foreign exchange markets just as its finance minister was in Washington, D.C., disclosing the country's inability to meet service charges on its massive $80 billion debt. "The World's Biggest Borrower Hasn't Got a Bean," promptly

Wishful Thinking and Bad Luck

reported the *Economist* in its August 21 issue. With September and a return to business at the world's banks and newspapers came series of articles that disclosed the magnitude of the problem that had been there all along. *Business Week* featured an article entitled "Worry at the World's Banks" in its September 6 issue. "Lenders' Jitters" was the *Wall Street Journal* headline on September 15. "Debtor's Prism," quipped the London *Economist* in its September 11 issue detailing problems in Mexico and Latin America. By October 16, the *Economist* had stepped up the rhetoric in an article entitled "The Crash of 198?" which discussed the Mexican crisis as one of a series including Drysdale, Ambrosiano, and Penn Square. Specific figures detailing the sum and distribution of developing countries' long-term debts emerged with amazing speed. More troubling figures on short-term debts were to come later.

Events of six months or a year do not cause a crisis like the one that emerged in August of 1982. They merely precipitate it. The causes are spread over much more time and, with diligence, can be traced back to events occurring decades earlier. The wishful thinking that resulted in tax cuts during 1981 and that in turn led to sharply higher projections of future U.S. deficits and the bad luck with monetary policy early in 1982 did not cause the debt crisis. They precipitated it, just as the unexpected closure of a lone gasoline station precipitates a crisis for the driver who, expecting the station to be open, has driven an extra fifty miles on a nearly empty tank. For the world economy and its financial system, the events of 1981 and 1982 came as shocks at a bad time. The tank was not empty when the crisis broke in August, but the banks and their borrowers were left running on their reserves. Another closed station would mean real trouble.

Chapter 11

Treating the Symptoms

Should I, after tea and cakes and ices,
Have the strength to force the moment
to its crisis?

T. S. ELIOT,
The Love Song of J. Alfred Prufrock

EVERY real crisis, whether it be a personal one confronting an individual or a worldwide calamity, passes through four stages: the first is denial, when people intimately involved in the crisis cling to the worldview that was held prior to onset; next comes panic when the old worldview evaporates faster than a new one can be constructed to replace it; third comes the confrontation stage during which we stare into the face of some adversity or adversary, taking its measure before acting to deal with it; finally comes the most painful stage of all, absorption, when retrospective denial, the distraction of panic, or the abstraction of prospective confrontation can no longer shield us from the fact that, knowing the world has changed profoundly in a way we had neither desired nor expected, we must live with it as it is.

Denial is, of course, the most dangerous phase of a crisis because by definition it precludes taking any corrective action to minimize its costs, while prolonged denial of a real crisis only increases the likelihood of debilitating panic. Panic is dangerous too because it usually results in either ineffective measures or, worse, counterproductive measures that only make matters deteriorate further. The banker who denies that his investments in developing countries are unsound cannot afford to stop pumping money into those countries lest his actions belie his words. If, when the interest stops coming in on the loans, he panics and lends the funds to pay the interest—as if to deny that there was anything wrong with the loans in the first place by demonstrating his faith in a turnaround—the more pathological is the initial denial of reality and the more likely it is that the panicky attempt to cover it up will preclude ever reaching viable confrontation and absorption stages.

Treating the Symptoms

A crisis can fail to progress beyond any of the stages prior to absorption; indeed the primary danger in a crisis is that the denial, panic, or even the confrontation phase will go on too long to permit effective movement toward the absorption phase. Neville Chamberlain never really got beyond the denial phase in dealing with Hitler, "taking his weekends in the country while Hitler took his countries in the weekend." The panic phase can end quickly and tragically, as it has for residents of high-rise buildings who try to escape a fire in their elevators. Some would say that the United States never got past the confrontation stage in Vietnam, ever-circling and assessing the foe without ever moving decisively to eliminate him or be eliminated trying.

Parties to the Debt Crisis Are at Different Stages

The parties to a crisis may pass through its different stages at different rates of speed, sometimes leading each other from stage to stage in a crisis being faced cooperatively, sometimes simply absorbing an opponent in an adversary crisis where one party moves far more quickly to the confrontation phase. The world debt crisis of the 1980s has, we have seen, four parties: the borrowing countries, the banks, the international organizations, and the governments of the developed countries whose banks are the heaviest lenders.

The borrowing countries have, by necessity at least, moved most quickly to the latter phases of the crisis. By August of 1982, Mexico had reached the confrontation stage, having no doubt been forced to abandon the denial stage when its bankers refused to roll over any more short-term loans. We must suppose that a period of panic, however brief, descended upon the top echelon of Mexico's government as it contemplated the collapse of its oil revenues together with the sharp increase in debt-service payments coming due in 1982. The emergence of Mexico "from the closet" of a repressed crisis had by early September released the floodgates of tension surrounding suppressed crises in Brazil, Venezuela, Argentina, Chile, and elsewhere, leading these countries quickly from the uncomfortable stage of denial, quickly through the panic stage—misery dearly loves company—and on to the confrontation stage.

Confrontation of the debt crisis by borrowers threw the bankers—who were unprepared to provide additional funding to the once-best customers for the loans they so ardently had been selling—into the denial phase, tinged with no little panic of the kind that produces a sickening feeling at the pit of one's stomach. The latter showed through in the September 1982 IMF World Bank meetings at Toronto as a sort of collective cloud of gloom that enveloped what is usually a kind of festival atmosphere. Explicit denials were issued by Citibank's ebullient chairman Wriston, who jauntily espoused the view—somehow repressed until September of 1982—that sovereign borrowers never went bust and never really paid back their debts. "Access to the marketplace"—the ability to keep borrowing (for a time at least, to provide the wherewithall to pay the interest charges on existing debts)—was the important thing for Wriston.

Some bankers probably took comfort from his remarks, although perhaps a little uneasily since his newly espoused worldview was not exactly the one that had led heads of the largest U.S. banks to commit twice their capital to a few politically volatile countries in Latin America. But if they had their doubts about Wriston's view of the world, they probably consoled themselves during much of the gloomy autumn of 1982 with the thought that the IMF and the Fed would come up with a plan—and the resources—to keep the big borrowers afloat. Such comforting thoughts evaporated when Jacques de Larosière introduced them to the IMF's bank "bail-in" concept at the New York Fed on November 16. Coming up with $5.0 billion in new money for Mexico to get the IMF to pitch in its $3.9 billion was not exactly what the bankers had had in mind. Elements of panic and confrontation swirled together in their minds, though the panic abated somewhat in December after successful implementation of the Mexican rescue package. But streaks of panic returned to the vortex of concern swirling about the persistent debt crisis after the new year presented—together with the realization that more Mexican debts would have to be rescheduled after just another year had passed—the dilemma of Brazil, even bigger than Mexico and sicker than had been thought at the initial diagnosis. Looming next to Brazil and Mexico were also Argentina, beset by impending sharp political change—historically the most reliable catalyst for default—and Venezuela, whose oil revenues were plummeting, not to mention basket cases like Poland and Rumania.

Americans at First Deny a Crisis

The U.S. government, and in particular the Treasury, was initially stuck at the denial stage of the debt crisis. At Toronto, Secretary Regan indicated no particular eagerness on the part of the United States to see the proposed increase in IMF quotas go forward, suggesting instead a counterproductively named "crisis fund" as a means to sidetrack the quota increase proposal by appearing to address a problem he really did not think existed. The U.S.

strategy, initially, was to try to cut the IMF out of a major role in dealing with the debt crisis, just as it had been cut out of a central role in managing major international financial problems during the 1960s and 1970s.

Like anybody who is busy denying the existence of a real crisis, the Americans were playing a dangerous game. Hobart Rowan, the *Washington Post*'s best plugged in writer on financial news, quoted a "senior U.S. official" on the eve of the meetings as saying that the proposal for a substantial increase in IMF resources was not warranted and has not been justified by its proponents. "What are they going to do with it [the quota increase]?" asked the senior official rhetorically. "We did not have that much of a [global] deficit in the first oil crisis or in 1979," said the official, referring to a proposal to double IMF quotas. Ever wily, Jacques de Larosière played along with the U.S. initiative for a crisis fund, telling Rowan—an always-reliable envoy to the Americans—that "anyone in the IMF who spits on the idea should be fired."[1] IMF staff members do not spit, especially when ordered not to by the "M.D." (managing director). But the Americans' initiative to sidetrack or minimize higher IMF quotas required them to minimize the debt problem and to characterize it as one that the banks, having got into it on their own, could get out of on their own.

Over the two and one-half months following the Toronto meetings, enough facts emerged about the gravity of Mexico's problems, not to mention those of Brazil, Argentina, and the rest of the heavy debtors, to convince the recalcitrant U.S. Treasury Secretary that maybe the debt problem was not just going to evaporate on its own. Emergence of a central role for the IMF as an honest broker between the banks and the borrowing countries—who else but the IMF could address the awkward problem that banks faced in bargaining with and extracting adjustments from their government clients?—had also added to the case for a quota increase. The U.S. "crisis fund" had long since fallen into oblivion.

In Paris on December 3 the United States announced that it had reached agreement with Britain, France, West Germany, and Japan to support a 50 percent increase in IMF quotas together with a $14 billion increase (from $6 billion to $20 billion) in the IMF's "heavy hitter club," the General Arrangement to Borrow that had heretofore been an emergency fund for the group of ten countries' own use. Under the new plan, its enhanced resources would be made available to all IMF members.[2]

The denial phase of the U.S. Treasury's reaction to the debt crisis turned to a combination of panic and confrontation in mid-December. Treasury Secretary Don Regan was testifying on Capitol Hill in favor of the administration's newly "essential" increase in IMF resources when Michigan Senator Don Riegal confronted him about giving money away to foreign countries while Americans were out of work. Riegal, the Democrat's official "big-business Republican baiter"—it was he who had orchestrated the revelation at the September confirmation hearing for a disgruntled Martin Feldstein that the chairman-designate of the President's Council of Economic Advisors was, like so many other high-level Reagan appointees, a millionaire with no idea of what it meant to worry about monthly bills for gas and electricity—reminded the unprepared Treasury Secretary that crisis denial can be dangerous. Over the fall of 1982 while the Reagan administration—with much encouragement from its most conservative elements—had been downplaying commitment of any resources to developing countries, the plain facts of the extensive exposure of the big U.S. banks in those countries had caused a rethinking of its position on the IMF quota increase.

But meanwhile the liberals in Congress were growing restive over the idea of what they saw as a bank bailout. The denial phase of the administration's approach to the debt crisis had consolidated conservative opposition to the IMF quota increase. At the same time it led the big banks—in private conversations with Treasury and their friends in Congress—into a near panic. These same banks had never before shown such interest in the resources going to the IMF, and their change of tune alerted congressional liberals to the political potential of a bank bailout issue. The big banks do not have many natural friends at the grass roots—recall the discussion of why they really only tolerate retail customers—and it was not very hard to mobilize a tremendous volume of mail against a bank bailout, especially since the banks were already busily running down their political capital with their ill-timed fight against the withholding of taxes from interest earnings.

The Treasury ended up caught in a vice on the IMF issue, squeezed between the conservative Republicans who hated the idea of sending more government aid to developing countries with their big governments, controlled markets, and absence of private enterprise and the liberal Democrats who were not about to support any proposal that helped the "fat cats" who owned and ran the big banks that—in their view—had been getting richer on high interest rates at the very time the poor and unemployed were suffering the most.

From Mexican Rescue to Brazilian Crisis

Just as the U.S. government was running into trouble on its newfound conviction about the need for more IMF funding, the December 15 deadline was approaching on Jacques de Larosière's Mexican rescheduling package. By

Treating the Symptoms

mid-December enough banks had—reluctantly in many cases, especially at some smaller U.S. banks and some British and Japanese banks—made their commitments to the $5.0 billion new lending package to convince the Fund to go ahead with its own $3.9 billion loan. The banks extracted an additional $800 million in fees and interest-rate premiums for their participation in the Mexican package.[3] The practice of adding to Mexico's debt burdens at the time it was least able to pay was widely criticized, but the banks saw themselves as needing compensation to undertake additional risks.

Completion of the first stage of the Mexican rescheduling package at the end of 1982 was something of a high point in the early stages of the debt crisis— a collective passage by banks, governments, borrowers, and the IMF squarely into the confrontation phase. The crisis had broken over Mexico with Mr. Silva-Herzog's dramatic August 13 visit to Washington, D.C. By moving quickly to the confrontation phase of its own crisis, Mexico had forced the world financial community to move quickly to confront its problem, as if it represented the key to solving the world debt crisis.

Retrospectively, it is clear that during the fall of 1982 the feeling among the banks and governments was that a successful Mexican rescheduling would put the crisis squarely on the back burner. When the crisis had broken in August, Mexico had erroneously been tabbed the world's largest borrower: "Mexico, the biggest borrower on international capital markets, has lurched in a few days from devaluation to exchange controls and then into the arms of the International Monetary Fund," wrote the cautious and usually well-informed *Economist* on August 21, 1982.[4] On September 6, *Business Week*'s coverage of the debt problem in an article entitled "Worry at the World's Banks" focused almost entirely on Mexico and reported data on U.S. banks with the most at risk in Mexico. (Estimated exposure totaled $13.5 billion for the ten largest U.S. banks.)

By mid-September more comprehensive numbers emerged that revealed that Brazil, with total debt of $87 billion compared to Mexico's $81 billion, was the world's largest debtor. But Brazil was labeled a model case among heavy borrowers. With a finger squarely pointed at Mexico as the bad boy of Latin debtors, the *Economist* singled out Brazil as an example that Mexico ought to have followed in language that must have produced some gnashing of teeth in Mexico City—not to mention some wan smiles in Rio and Brasilia.

Mexico would have done far better to have followed the path taken by Brazil in 1980. Brazil was then the world's biggest debtor and looked a Mexican-style crisis between the eyes. With no help from the IMF, within a year it turned a trade deficit into surplus by curtailing industrial growth, increasing prices to its farmers and devaluing its currency. Brazil continued to borrow from foreign banks but avoided getting repayments bunched by conceding the higher rates asked on long term credits.[5]

The *Economist*'s glowing report on Brazil was not an isolated case. In a lead article entitled "Lenders' Jitters" in its September 15, 1982 edition, the *Wall Street Journal* did not get around to discussing Brazil until the end: "Brazil, the other major depository of U.S. lending, already has taken some serious economic overhaul measures and isn't regarded as an immediate problem."[6] The *Journal* went on to qualify its optimism with the suggestion that some observers saw political problems arising in Brazil in response to what were the government's attempts at austerity. Brazil had not sought IMF aid but neither was it pursuing any vigorous austerity measures, as would soon become clear when it did seek the help of the IMF.

One year after its article extolling Brazil and chastizing Mexico, the *Economist* reported on September 3, 1983, a reversal of roles in an article entitled "A Mexican Model for Latin American Debt."

There was almost a standing ovation at last week's ceremonies in New York which formally rescheduled $11.4 billion of Mexico's foreign debt. Foreign bankers lining up to sign eight-year credit extensions had nothing but admiration for the way Mexico's new administration had implemented the International Monetary Fund's stiff austerity programme and yet had managed somehow to avoid serious political unrest.[7]

Under the heading "Black mark for Brazil" in the same article, the *Economist* reported on the struggle between the IMF and Brazil over its nonfulfillment of terms on its IMF loan.

The sharp reversal of roles for Mexico and Brazil that occurred between the Septembers of 1982 and 1983 was symptomatic of a new phase of the debt crisis. The obsession with Mexico during the last four months of 1982 was a dual-faceted form of crisis denial. It was constructive in that it concentrated resources on a problem of manageable scale, but destructive insofar as it created the impression that Mexico represented the key to solving the debt crisis. Like the victim of an attack by a gang of thugs who subdues one attacker as if that will make the others run away, the governments, banks, the IMF, and Mexico wrestled the Mexican debt crisis to the ground in the closing months of 1982. When they stood up on wobbly legs, exhausted by weeks of intense negotiation and ceaseless travel, the rest of the gang closed in.

Treating the Symptoms

Brazilian Nightmare

On December 20, 1982, Carlos Langoni, the young, bespectacled governor of Brazil's central bank—who a little over eight months later was to resign his post in protest over unrealizability of stringent IMF terms on loans to Brazil —addressed a large group of nervous bankers at New York's Plaza Hotel where, according to Plaza advertisements, "nothing unimportant happens." Mr. Langoni explained to the bankers that Brazil, the apple of their collective eye, could not pay its debts coming due in 1983. Plus, he added in a now familiar one-liner that follows such bad-news revelations by big debtors, Brazil wanted $4.4 billion in new money together with maintenance of credit lines to Brazilian importers and the foreign branches of Brazilian banks. The latter condition, as the bankers were soon to discover much to their dismay, referred, obliquely at the time, to roughly $10 billion that Brazil, with some effort at concealment, had been rolling over in the market for overnight funds (borrowing and reborrowing from day to day). Flanking Mr. Langoni was the ubiquitous IMF chief, Jacques de Larosière. He was there to explain policies that the IMF saw as necessary to overcome Brazil's difficulties, signaling clearly the accession by the Brazilians—who like the Mexicans before them had long eschewed IMF involvement in their affairs—to the need to endure the humiliation of an IMF loan and the strings attached to it. In exchange for the banks' granting Brazil's request for additional funding—following a pattern similar to the one established for the Mexican package—the IMF was prepared to provide Brazil with a three-year $4.9 billion standby loan agreement, provided that it could produce a 1983 trade surplus of $6.0 billion and a significant reduction in its horrendous inflation rate of well over 100 percent a year. The latter condition was the toughest in view of Brazil's 30 percent devaluation in February of 1983—itself inflationary—and of the guaranteed self-perpetuating wage-price spiral that was built into the indexation of Brazilian wages and salaries to automatically adjust upward at rates equal to or greater than the inflation rate.[8]

The Mexican model initially provided a compelling example to follow, although in the Brazilian case, some pretty diligent wishful thinking was required. By February Brazil had signed an agreement with the IMF for the $4.9 billion with the first disbursement earmarked to repay Brazil's $1 billion-plus loan from the BIS coming due in July. Beyond the IMF funds, Brazil got a bank loan of $4.9 billion, rescheduling to longer maturity of $4.7 billion in amortization credits for 1983, and restoration of interbank credits to $7.5 billion together with $9 billion in short-term trade credit. The "new money" part of the package was the three-year, $4.9 billion bank loan. The rest was a commitment to continue existing credits that had been effectively suspended pending arrangement of the February package.

By May it was clear that the February package was not going to work. The IMF staff signaled its dissatisfaction with the continued acceleration of inflation together with the lack of progress in reducing Brazil's own public sector debt and withheld dispersal of the first $1 billion of the IMF standby loan that would have gone to repay Brazil's BIS loan.

Failure of the First Brazilian Package

The collapse of the February 1983 package reflected the fundamental problems that banks and governments had been failing to confront, first in concentrating on the Mexican package alone and then by slapping together a Brazilian package that had no hope of succeeding. The spring 1983 unraveling of the Brazilian package, together with emergence of further problems in Argentina and Venezuela (to mention only the most prominent cases), brought the banks and governments face to face—probably for the first time—with the full magnitude of the world debt problem. The nine months of struggle from September of 1982 to May of 1983 brought the bankers squarely up against a reality and immensity of the problem that had simply not been comprehended. Some panicked and some confronted it.

The basic reasons for failure of the February 1983 Brazilian rescue package lay in the retreat of both borrowers and lenders from commitments they had made. And the movements of each were self-reinforcing. As the months passed after February it became clear that, while Brazil was making some progress toward its goal of a $6 billion trade surplus—largely by throttling imports and pushing exports—its inflation rate and public sector debt were soaring. Like Jimmy Carter, but on a far greater and more tragic scale, Antonio Delfim Netto, Brazil's rotund, fast-talking minister of planning, was trying to create growth just by pushing up demand. Ironically, Delfim Netto's program was called, like Nixon's a decade earlier, a New Economic Policy designed to "fight inflation with growth." But Brazil's internal government deficit, with spending running way ahead of tax receipts, was rising, and money was being printed to finance such deficits. The result was ever-faster inflation that Delfim Netto answered with an incomes policy pegging wage increases above inflation rates. The result was a further speed-up of inflation.

Treating the Symptoms

Under the terms of the February letter of intent signed along with the IMF package, these policies were to have been scaled back, with wage increases set below inflation rates and slower money growth. This simply was not done, and it began to look to the banks like Delfim Netto's strategy was to draw them in even further so that heavier sunk costs would make it even harder to get out of Brazil—no matter how badly it acted. Brazil was weak but as the banks kept pumping in money, they too became weak, closely tied as they had become to Brazil's fortunes. The big banks—Citibank, Bank of America, Chase, Manufacturers Hanover, Morgan, and Chemical—just could not ignore the fact that their exposure in Brazil alone averaged 65 percent of their collective equity and that a Brazilian collapse or default would trigger the same for Argentina, Venezuela, Chile, and perhaps even Mexico—whose debts accounted for twice their combined equity.

The tension between banks over Brazil began to show in February as Bankers Trust, which was trying to hold together the interbank network lending to that beleaguered country, began to encounter heavy resistance from regional banks that were being asked to maintain commitments to Brazil and elsewhere in Latin America. The banks' share of the overall Brazilian package was being managed largely by Morgan and Citibank. The dominant personality during the spring of 1983's rescue effort was Morgan senior vice president Antonio Gebauer, aggressive, fluent in Portuguese, the architect of Morgan's presence in Brazil, and described as "the epitome of the petrodollar recycling mentality."[9] Gebauer apparently saw himself inextricably tied to the success or failure of the Brazilian package and was accused of trying to "browbeat" reluctant banks with a lesser stake in Brazil into participation in the rescue effort. And as many bankers saw the inflation, money supply, and budget deficit numbers on Brazil's roll in during the spring of 1983, they wanted less and less to do with that country, especially if they were not already too heavily committed to cut and run.

Confrontation—Phase Two

By late May the Brazilian crisis was squarely into the confrontation phase. Moreover, since Brazil was so huge and no one could ignore that serious problems in Argentina and Venezuela were waiting in the wings—the entire debt crisis was coming to a head. The bankers were talking about Brazil but they knew that was just code. They were actually talking about the whole towering superstructure of some $200 billion in Latin American debt. It did not matter that this was far less than the total $700 billion in LDC debt. A collapse of $100 billion of the total wherever it occurred would be more than enough to wipe out the capital of the biggest U.S. banks. No one cared to think hard about the fallout from that calamity, either in terms of consequences for themselves and their banks or in terms of consequences for the world economy.

Phase two of the Brazilian crisis—the world debt crisis—came to a head on May 31, when Paul Volcker called the heads of the big banks to the New York Fed for a meeting with himself and Tony Solomon, Carter's Treasury Undersecretary, now President of the New York Fed. Volcker needed only ask, "What are you going to do about the problem?" (unraveling of the Brazilian package) to signal the nervous bankers that he felt some changes were in order. Gebauer was replaced by Citibank's sharpest debt renegotiator, William Rhodes, who had until then been totally preoccupied with Mexico and Argentina. Rhodes, the best in the business, moved decisively to organize his efforts, keeping in close contact with Messrs. Volcker and de Larosière, both of whom he was comfortable calling at any time he felt the need. Meanwhile the IMF had held up disbursement of the $1 billion-plus first distribution of its $4.9 billion standby loan in light of Brazil's failure to perform satisfactorily in meeting conditions that had been imposed. This in turn put the banks on hold. They were not about to distribute any more funds to Brazil until the IMF was getting compliance. The resulting freeze on the funds flowing to Brazil meant that it was unable to repay its $1 billion loan from the BIS when it came due in July; nor was it able to make regular interest payments on its $90 billion in debt. Arrearages began to mount, reaching $2 billion by August.

American Politics

As phase two of the Brazilian crisis heated up, the debate in the U.S. Congress over U.S. support for the IMF quota increase—off to a bad start with Secretary Regan's bungled testimony the previous December—began to heat up as well. The conservative-liberal coalition that had grown up in Congress opposed to funding the U.S. share of the increase—which would add an $8.4 billion U.S. commitment to an overall package about five times as large to IMF resources —was sizable enough to pose a serious threat to passage of the IMF bill. And passage of the bill was crucial for a number of reasons. Although the U.S.

contribution was only about 20 percent of the total package, it constituted about twice as large a share of the usable currencies in the package since many of the Fund's poorer member countries would be supplying their quota increase with essentially worthless paper currency. Second, U.S. participation was the key to an additional $6.0 billion special package being put together in about equal parts by the Saudis and the European countries. This made the U.S. IMF quota increase the key to almost $15 billion in additional funds to try to carry the floundering Latins through the crisis. It was easy to predict in the summer of 1983 that the funds from the quota increase together with the additional $6 billion Saudi-Euro package, if added to $6.5 billion in new money from the banks—up from February's $4.4 billion because of bigger Brazilian needs and bolting by some banks—would be barely sufficient to enable Brazil just to service rescheduled debt through the end of 1984. The BIS, as Fritz Leutwiler had clearly indicated in his August 20 outburst, was not about to commit more funds.

Liberal congressional opponents of the IMF "bailout" of the "big" (read greedy) banks had gained considerable support from displeasure over the heavy lobbying efforts expended by the banks in the spring to defeat the interest-withholding law. The conservatives still did not like the idea of pouring more money into government-run economies—especially in view of the revelations of waste and corruption in Brazil that were beginning to emerge together with scenes of rioting in the streets by opponents of IMF-mandated austerity programs. The President and White House staff—having been lulled into a state of quasi-complacency by an April report, National Security Directive Three, which concluded that "the medium-term prospects for improvement in the debtors' financial position are quite good"—were not pushing very hard to keep the conservative Republicans in line on what was after all the administration's proposal for quota increase funding.[10]

Matters deteriorated further when, after a heavily amended House version of the IMF bill had cleared Congress, the National Republican Congressional Committee issued a news release that attacked twenty Democrats for having opposed an amendment that would bar the United States from supporting IMF loans to "communist dictatorships." The renegade conservative Republicans further needled the Democrats, charging that they had "voted . . . to loan U.S. taxpayers' money to communist nations."[11] The whole thing was obviously a planned setup since Republican Congressman Phil Gramm of Texas had attached to the bill the amendment requiring Dick Erb, the U.S. representative on the IMF's executive board and now the IMF Deputy Managing Director, to vote against loans to any countries under "communist dictatorships." Besides the Gramm amendment, liberal amendments attached bank-punishing terms to the bill that effectively made it far more costly for banks to extend further loans to Brazil and other ailing LDCs.

Congress Fiddles While the Financial System Totters

The spectacle of a U.S. Congress playing domestic politics with IMF quota legislation while a strong President diligently looked the other way and Brazil's arrearages mounted toward $4.0 billion in the early fall of 1983 suggested to the rest of the world that the Americans were either ignorant of the crisis their banks were facing—not likely in view of the well-organized bank lobby—or simply exhibiting an extreme case of the shortsighted parochialism on international problems that has characterized U.S. leaders from FDR putting "first things (us) first" to Nixon's not giving "a shit about the lira." The fallout from Republican accusations on the Gramm amendment jeopardized final passage of the IMF bill at the very time when concern over the world financial system was at a frenzy peak. On Sunday, August 14, 1983, no doubt not coincidentally just one year and a day after Silva-Herzog had fired the opening shot in the crisis, Eliot Janeway, under a heading "The Prince of Pessimism Says the System Could Topple," disturbed a quiet breakfast for more than a few lazy summer readers of the *Washington Post* with a chilling diagnosis:

> The west's bankers and politicians are misleading the public with their assurances that the huge load of short-term international debt constitutes a manageable problem. The fact is that much of the interest on these loans cannot be paid, and most of the principle will not be.
>
> The situation is spinning out of control and poses a clear and present danger for the entire international financial system comparable to the one preceding 1929. Only drastic action by Western governments impelled by a realistic sense of urgency will head it off.
>
> The theater provided by the House the week before last, when it narrowly approved a complicated transfusion of up to $8.4 billion for the International Monetary Fund, advertises the problem rather than solves it. A band-aid is being prepared for application with full ceremonial flourish, but the hemorrhage continues.
>
> The proposed $32 billion increase to be raised from all the IMF's members would buy time. But it would also nurture the dangerous illusion that we have time. In fact, it comes too late and offers too little to solve the underlying problem.
>
> New short-term loans to keep interest payments rolling into Western banks will not revive the world economy. They will not start cargos of goods moving between ports, or finance productive new investment. They will only pile debt upon debt, most of it unpayable.

Treating the Symptoms

I know I have been described on Wall Street as the "prince of pessimism," but I prefer to describe myself as an "early bird alarmist"—an optimist who believes solutions are possible. Promises that the crisis can be handled with business-as-usual methods do not help. People who make them are not doing simple arithmetic. The debt of developing countries and communist Eastern Europe is now in the neighborhood of $750 billion. At 14 percent, the interest on that float comes to around $105 billion a year. Put another way, bank interest compounded on unrepayable principle will double the debt load in five years, even without adding in the heavy charges and fees that the Western banks have been "persuading" their debtors to accept in the latest rounds of refinancing.

The charade has crossed new frontiers of creative accounting. Banks advance the interest owed them with one hand and take it back with the other. No real money changes hand. It's a numbers game.

The realistic definition of bankruptcy is the point at which a debtor can no longer borrow the interest owed. By that stern test Poland and Brazil already are cold, stone broke. One official default is all it will take to stampede the other busted borrowers into a full-fledged cartel of defaulting debtors.[12]

Janeway put it squarely on the line. All the scrambling for more funds for Mexico, Brazil, and the other Latins—not to mention virtually hopeless cases like Poland and Rumania—had nothing to do with the banks' getting repaid. Rather it was only a scramble to hand over to these mega-borrowers the wherewithal to enable them to hand right back the very same funds that would then be labeled interest payments on their mega-loans. And the regulations say that as long as interest payments do not get over ninety days in arrears, the loans need not be characterized as "nonperforming"—a term that implies, first, a subtraction of six months' interest on the loan from bank earnings. If Brazil was dubbed nonperforming, it could eradicate almost half of Citibank's 1983 earnings together with sizable chunks of other big banks' earnings. Of course, irremedial nonperformance—as Janeway suggests, the result of chronic inability to borrow interest owed—means bankruptcy for borrowers and default for lenders. And we have already discussed what default would mean.

Reagan Responds

During the late summer the Treasury and White House apparently joined those who were ruminating about the prospects for the world financial system and, closer to home, were beginning to worry about the outlook for the large banks and the smart recovery of the U.S. economy that was underway. In his September 27 address to the Joint Annual Meetings of the IMF and World Bank at the Washington Sheraton, President Reagan strongly endorsed the IMF as the "lynchpin" of the world financial system. Delegates had expected the President to support the IMF bill, but after the warm applause the murmur of approval among delegates leaving the Sheraton Grand Ballroom was punctuated with phrases like "stronger support than expected" and "heartening commitment."

In reality, the moment was a very significant one. In his usual upbeat manner the President himself had acknowledged existence of the crisis by displaying stronger support for an IMF that only a year earlier he had been disinclined to support. The President's progress to the confrontation stage of the crisis made it very likely that sufficient pressure would be applied to extract an IMF bill from the joint House-Senate committee that had still to hammer out a compromise version. Mr. Reagan's progression also moved everyone—including the banks, the borrowing countries, and the IMF—closer to the painful day when the reality of the crisis would begin to be absorbed and managed.

The President's warm, strong support for the IMF—which he characterized as "the Dutch Uncle, talking frankly, telling those of us in government things we need to hear but would rather not"—was the high point of the otherwise grim annual meetings. A year earlier all the bankers and finance ministers had gathered at Toronto to find almost everyone there in a state of shock; some, like Wriston and the U.S. delegation, had denied anything was really wrong, while others felt panic. A year later in Washington the shock had worn off and there was no sign of real panic. Intense emotion had given way to the dull realization that there were billions in assets on bank balance sheets—not to mention billions in unpaid interest already included in bank earnings statements—that were not worth anything near their book value, the 100 cents on the dollar at which they were currently counted. The crisis had turned out to be far worse than anyone at Toronto, obsessed with Mexico, had imagined. In September of 1982 Brazil was still the apple of every banker's eye. The bankers who plied the chrome-and-velvet, new-upbeat-hotel-look Sheraton lobby were in a state of mind akin to that of out-islanders who had just weathered a tidal wave. They had survived but the forecast called for more tidal waves, possibly bigger than the first, and they could not be sure that the next wave—Brazil—or the one after that—Argentina, or Venezuela, or the Philippines—would not sweep them away.

Treating the Symptoms

Political Risks Compound Economic Risks

The seven weeks between the end of the Washington-Sheraton meetings and the adjournment of the U.S. Congress on the Friday before Thanksgiving were full of tension and drama. Any illusions that might have been carried away from Washington about a period of calm were promptly smashed when Julio Gonzalez del Solar, affable and easygoing sixty-six-year-old president of the Argentine central bank, was arrested as he stepped off a commercial airliner on his return from the IMF annual meetings. Federico Pinto Kramer, a district judge, in the remote Argentine town of Rio Gallegos, had ordered his arrest and questioning on the charge that del Solar had failed to represent Argentina's interests in the renegotiation of part of its $40 billion in foreign debts —the rescheduling of $220 million in debts of Aerolineas Argentinas, the state airline.

This bizarre incident, perpetrated by an obscure judge known for his links with ultranationalists, snapped into focus the realization—already taking shape from media reports of Rio citizens looting supermarkets and staging massive protests over government austerity measures—that foreign debts were an explosive political issue in addition to being an economic nightmare. The judge's rationale for the arrest, that there were irregularities in the way much of Argentina's foreign debt had been incurred under the military government, provided a chilling reminder to those familiar with the history of repudiation —the invalidation of "treaties of tyrants" by governments that take power after "oppressive" regimes and renounce, amid cheers of the citizenry, the "illegitimate" acts of their predecessors, including especially the debts they have contracted.

The discussion of Argentine repudiation was by no means hypothetical in view of the fact that a hotly contested election to select a successor government to Argentina's discredited unpopular military regime was less than four weeks away when news of del Solar's arrest hit world headlines. A report on Argentina in *Business Week* caught the October mood:

> The international debt crisis is fast slipping out of the hands of governments and commercial banks—and right into the supercharged world of Latin American politics. Just days after the International Monetary Fund and the World Bank assured the world that the global debt problem would become more manageable in coming months, the lid blew off.
>
> In the remote Argentine town of Rio Gallegos, a district judge may have started a potentially explosive political backlash against IMF austerity measures. He ordered the arrest and questioning of central bank President Julio Gonzalez del Solar

about a debt rescheduling package for Argentina's national airline, Aerolineas Argentinas. The judge charged that its terms were unconstitutional.

Even if he is eventually overruled, the incident revealed an ominous upsurge of populist and ultranationalist political resentment against foreign creditors that is also building in Brazil and other major debtor countries. If it continues to gather momentum, the odds are strong that the Argentine government will eventually be forced to repudiate or declare an extended moratorium on payment of all or part of its $40 billion debt—$8.2 billion of which is owed to U.S. banks—and that several other hard-pressed countries will be encouraged to follow. Such a decision would throw international money markets into turmoil and make it even more difficult to nurture the fragile world economic recovery into a stronger one.

"No Leadership." The sudden turn of events caught Washington officials and the major banks by surprise. Only days ago there were wide expectations that the next 12 months would show both borrowers and lenders that there was light at the end of the debt tunnel. Now, an anxious Reagan Administration official warns, "there is no leadership on the financial side other than 'let the markets take care of this.' " Everyone in the banking community, adds a U.S. banker, "is scared." After spending 14 months patching over crises in Mexico and Brazil, experts now concede that Argentina will be the critical test of their ability to manage the debt mess without a major financial catastrophe.

A debt rollover for Argentina's airline, negotiated by Morgan Guaranty Trust Co., was to be the model for refinancing of up to $10 billion of other government debt. After the district court blocked it, Buenos Aires' military government, in a desperate attempt to preserve precious hard-currency reserves, suspended the automatic disbursement of foreign exchange. Foreign banks immediately shut off all loan payments and trade financing to the country. The unfolding drama sent shock waves through the international financial system.

Aggravating it all is the political agitation in Argentina generated by the campaign for national elections on Oct. 30 to replace the discredited military junta. Local and foreign bankers in Buenos Aires have little faith in the pledge by President Reynaldo Bignone, head of the outgoing military regime, that Argentina "will honor" its debts. More significant, they fear, is the political attitude expressed by Saúl Ubaldini, general secretary of a powerful Peronist labor confederation, who charges that in its handling of the debt, the junta "surrendered totally and absolutely" the patrimony of the nation. Any new civilian leadership is expected to be weak and vulnerable to such political forces.

Already international banks had halted disbursements of $500 million and the IMF is appraising the situation to determine whether to suspend its program for Argentina. This would deprive Buenos Aires of another $900 million. Argentina has stopped paying for all but essential imports.

"If there is a country that thinks it can try and get away with a repudiation, it's Argentina," says a U.S. banker. This attitude, he explains, stems from the country's abundance of export products: meat, grain, and soybeans. Adding to Argentine confidence is its near self-sufficiency in energy. If Buenos Aires takes the plunge, observers fear that Venezuela and Peru might follow suit. Even if they do not, bankers concede that nearly all other major debtor nations will demand easing of their debt burdens.

To forestall such a development, major international banks were already preparing to soften loan terms for several major Latin debtors. These would probably take the form of longer maturities, lower rescheduling fees, and reduced interest charges on existing debt. Some loans, concede bankers, could be carried at below-market interest rates. By making these concessions, the bankers hope to salvage at least part of the principal on their loans. The alternative is to simply write them off. "There is no way you can take this old debt and consider it something to be paid off at current market rates," concedes the executive vice-president of a major U.S. bank. "That was the premise of the original Mexican, Argentine, and Brazilian reschedulings."

Unattainable Targets. Such concessions to borrowers will be reflected, analysts say, in lower bank stock prices. And a U.S. official fears that the banks will try to recoup the revenue domestically, "through charging higher rates on US corporate and consumer loans."

But bank concessions will not be enough. The IMF, say most observers, will also have to ease the harsh economic adjustments it has demanded of borrowers, often setting unattainable economic targets as a condition for its own loans. So far there is little sign of such softening either on the part of the IMF or in the thinking of the Reagan Administration, which has taken the lead in demanding tough conditions for debtors.

Furthermore, "each unraveling [of the debt rescue net] is going to lead to an increased reluctance of banks to lend," points out a U.S. official. Indeed, growing numbers of U.S. regional and European banks have refused to maintain interbank credit lines or participate in new loans to Third World debtors. And several big U.S. banks are already at their legal lending limits to some Latin countries. As the crisis drags on, Washington and bank officials fear that the U.S. banking system will become isolated from its European and Japanese counterparts. Says a ranking U.S. banker: "The Europeans will say, 'Latin America is an American problem.' "[13]

Boiling along with news over political turmoil in Argentina were troublesome reports on Brazil and the fate of the IMF quota increase. The Brazilian Congress rejected by a vote of 267 to 1 the IMF-mandated wage indexation scheme, a condition for its Brazilian rescue package to go forward. The Democrats in the U.S. Congress refused to support the IMF quota legislation unless the President himself apologized to them each, in writing, for the "supporting communist dictatorship" charge of the National Republican Congressional Committee. For good measure, Representative Ferdinand St. Germain (D-R.I.), wily, tough chairman of the House Banking Committee, tacked on a $15.6 billion housing reauthorization fund to put the government back into production of housing for the poor as a condition for his support, without which the IMF bill could not pass. Brazil's debt repayments fell more than $4.0 billion in arrears. Added to these concerns were reports that time was running out for Venezuela on refinancing of $18.4 billion of its total $34 billion in foreign debts. Over $25 billion of Venezuelan debt was due by the end of 1984. Like Argentina and Brazil, Venezuela was balking at adoption of an IMF adjustment program. In addition, Venezuela was looking ahead to an election in December.

Buying Time

The wavering international financial system regained some of its footing in November with a rapid-fire succession of major coups, each of which was necessary for avoidance as a total collapse. After the President wrote a letter commending the courage of Democrats who had voted for the early version of the IMF bill, the log-jam broke with passage by the U.S. House and Senate of a compromise IMF bill that included some wrist-slaps for the banks and Congressman St. Germain's $15.4 billion for housing for the poor—two dollars of aid for U.S. poor for every dollar doled out to foreigners.[14] The Brazilian Congress finally approved a watered-down wage indexation scheme, and the IMF executive board voted to resume lending to Brazil. Brazil was thereby entitled to draw $1.2 billion in 1983 as part of the three-year, $4.5 billion IMF loan. The country used the funds to pay overdue emergency rescue loans from the BIS. The IMF action freed up $8.5 billion in loans from commercial banks —$2 billion in 1983 loans that had been suspended and a possible $6.5 billion in new loans.

On October 30 the Argentinians elected as president Raul Alfonsin, of the left-center Union Civica Radical, a surprise winner over the fiercely nationalistic, populist Peronist party candidate. Both candidates had promised to honor all "legitimate" foreign debts, hardly a consoling pledge in view of election rhetoric about illegitimacy of the military government and Alfonsin's pledge to pursue perpetrators of human rights' violations among the outgoing generals who had ruled Argentina with a heavy hand since ousting Isabel Peron in 1976.

Jacques de Larosière journeyed to Basel on November 10 to shore up the commitment of noncommunist central banks—excluding the U.S. Federal Reserve—and the Saudis for a $6.3 billion standby credit from which the IMF could draw over a two-and-one-half- to three-and-one-half-year period. After the U.S. Congress had acted a week later, the BIS formally approved the $6.3 billion credit at its December meeting. That credit, together with the GAB increase of about $12 billion and the $18 billion or so usable currency portion of the IMF's $30 billion-plus quota increase, brought Fund resources available over the three years following the end of 1983 up by about $36 billion, at least a third of which was already committed.

Early in December after all this had come to pass, the players in the debt crisis drama could pause for a breath of air. Like soldiers traversing a minefield, they had done every difficult thing needed to reach the other side. Success had been improbable but they had made it anyway. But "success" only meant staying alive to fight another day. The debt crisis players were beginning to feel like bone-tired and battle-weary troops who begin to question the reason for fighting when victory only means that they win the chance to fight again. When generals on both sides of a war begin to sense such feelings in their soldiers, perhaps the time has come to think of counting up losses and negotiating for a truce that defines what is left for each side to hold on to.

Beginning to Absorb the Losses

By the end of 1983 the major result of the interim resolution of the IMF quota-increase spectacle and the nth rescheduling package for Brazil was the dawning of the realization that the world debt crisis was not going to peak rapidly and then just go away. It was taking on attributes of a chronic ache, so painful that it keeps sapping strength from its victim. The bankers were said to have learned something from the IMF-Brazil phase of the crisis. *Business Week* reported that the bankers "have become keenly aware that the rates charged Brazil and other debtors made it almost impossible for those nations to meet interest payments, creating the threat of an international banking collapse."[15] That was a rather odd way of saying that Brazil—which had eagerly agreed two years earlier to contracts specifying just the rates it was being charged—could not pay the interest on its debt unless the banks lent it the money to do so. Viewed in this light, the concession whereby banks cut their lending rate from 2.5 to 2 percent over LIBOR meant only that the banks would have to come up with a little less money to hand over to Brazil as the means for that nation to hand back interest payments. The banks' princely concession was estimated to save Brazil $15 million a year—just under 12 hours' worth of interest payments in 1984. The banks also took fee cuts of $60 million, a little more generous but hardly likely to change much.

In Congress it was clear that a mood for much tighter bank regulations—counter to the extant strong tide of domestic deregulation—was swirling around in the debate on the IMF bill. As it emerged, that bill contained some congressional mandates: special noninterest-earning reserves to cover possible overseas loan losses; requirement of full disclosure of exposure to foreign country debt on a quarterly basis and not just disclosure of problem loans; and an ominous blanket requirement of "adequate capital levels" meaning that a watering of shareholders' equity would not be permitted as a means to control bank losses.

The banks, prodded by the enormity of the debt problem and their dreadful public image—which became painfully vivid during debate on the IMF bill—were beginning, very tentatively, to explore graceful ways—or at least graceful-sounding ways—to absorb some of their heavy losses. Most desirable would be measures—such as accepting lower interest rates on new and existing loans —that would at least help to shore up their public image as greedy, insensitive bloodsuckers while acknowledging reality. No doubt unemphasized would be the fact that if you have a one-dollar loan earning, say, one percent over the market (London interbank) rate—about 14 percent—a reduction to a 12 percent interest rate cuts the value of the loan to about 12/14 of a dollar, or to about 86 cents. A one-dollar contract that calls for you to receive 14 cents a year is not worth a dollar if you only get 12 cents a year. The banks know this acutely, and in time so will all of their shareholders.

Another end-1983 sign of the permanence of the debt crisis was its wide-spread discussion outside of traditional banking and financial circles. *Rolling Stone* magazine's William Greider wrote "Uncle Sucker," terming the debt crisis "a very dicey game" and asking why the United States should "bail out big banks for flakey loans to defaulting Third World nations."[16] Small newspapers that usually avoid heavy financial stories like the plague carried reprints of the *Rolling Stone* piece—as if its appearance had signaled to the followers in the printed media the legitimacy of the debt crisis as a bona fide topic for filler in the "thought-piece" sections of the Sunday edition.

Ever since late 1982 newspapers and magazines had been full of cartoons depicting bankers acting like spendthrift fools on international loans while treating individual borrowers with the needlenose contempt of a Shylock or a Scrooge for his borrower-victims. In all the media, economic crystal-ball gazers conditioned their forecasts on the scary implications of higher interest rates or an aborted recovery for the "international debt problem," as it came to be coded to distinguish it from our own (rising) debt problem of huge government deficits.

The debt crisis was becoming a media event because it met the two criteria: First, it was big enough to catch audience attention by suggesting that it might affect "*you*, Mr. and Mrs. Average American"—will an international collapse cause your bank to fail and/or mean another Great Depression? Besides, being big, the international debt crisis was no flash in the pan and it had drama.

Treating the Symptoms

Congress debated it hotly, statesmen and central bankers discussed it at high-level meetings and made public comments in somber tones. Best of all, the whole thing had the big-city bankers in their striped suits looking like a bunch of fools. That's box office.

But the debt crisis was different. It was not going to be like a space shot or the Olympics. "The thrill of victory or the agony of defeat" would not reach us in a flickering image on television. Nor was there much prospect for some sort of collective triumph like a moon shuttle. Unfolding, it was dramatic and sometimes exciting; but it was essentially tragic. December of 1983 was not, however, a time when most Americans—on the first real Christmas shopping spree in years—spent much time thinking about the debt crisis. Why should they? There was nothing much they could do to help Brazil undo years of overspending on too many government workers or hydroelectric dams that wouldn't really be needed until the year 2000. Nor could many of them expect to affect the outcome of the battle that was looming to decide just who—the banks, the taxpayers, or both—was going to eat the $15 billion or so in loan losses that the marketplace was saying were on the books of the big U.S. banks that collectively earned "only" about $4.0 billion in 1983.

Chapter 12

For Whom the Bell Tolls

A national debt, if it is not excessive, will be to us
a national blessing.

ALEXANDER HAMILTON,
Letter to Robert Morris, April 30, 1781

THE LDC debt crisis itself is not a very subtle matter. A group of large banks, with a disproportionately high representation of U.S.-based banks, were spurred on during a commodity boom to overlend to LDCs. The process has been repeated countless times, as is ably recounted in Charles P. Kindleberger's lively account *Manias, Panics and Crashes.*[1]

As does every crisis, this one has its unique aspects. Most significant is the fact that this time, instead of acting only as intermediaries to distribute the paper claims of what were ultimately to become distressed borrowers, the banks ended up holding those claims in their own portfolios—"holding the bag," as it were. As a result, this particular debt crisis threatens not only the liquidity of the lenders intimately involved but the liquidity—and perhaps the solvency if handled carelessly—of the *world's* financial system, for the banks are at the heart of that system. In short, the banks have not only encouraged a speculative binge—that is, after all, their usual blameless role in such episodes as providers of credit to speculators—but this time they themselves have been the speculators. The result of this outcome is an awkward situation for governments and potential lenders of last resort—not to mention the banks and LDCs themselves—that must disentangle, or at least defuse, the crisis.

The really interesting thing about this crisis, with its common and unique qualities, is the tremendously broad and seemingly disjointed set of circumstances that had to combine to bring it about. Before the fact those circumstances swirl about, extremely unlikely to combine in a coherent way for good or evil; after the fact there they are, once highly improbable and now seemingly inevitable.

These disjointed yet compelling sources of the LDC debt crisis have been the subjects of this book. First there were the fears of industrial societies about raw materials shortages, reinforced at just the right time in 1972 by the Club of Rome's *The Limits to Growth*[2] and then realized with the first oil crisis. On the other side were the huge government borrowing consortia in LDCs that reflected their fears of outside control. The consortia imposed themselves between individual projects and the banks, appearing to the banks as "sovereign" protectors of the banks' interests while enabling them to take hugely profitable hundreds on millions in loans at a clip.

Behind all this lay the ongoing development of a true world debt crisis of which the LDC debt crisis was only an acute phase. Since the 1960s governments had promised social programs, the realization of which ought to have been conditioned on the rosy economic assumptions that eased their adoption. Less growth and the U.S. "guns and butter" years meant that by 1971, Richard Nixon had to break away from the crude but effective discipline of gold. The result was a worldwide inflation surge that preordained the success of the OPEC cartel.

Coupled to the general explosion of government debt were the problems with the United States in its ambivalent role as world economic leader. It rejected the role in the interwar period, just when the British could no longer fulfill it. It snatched world financial hegemony from the IMF in 1945, cementing its role until 1960 with piles of gold, hard currency, and Marshall Plan largesse. Attempts to prolong dollar-centricity at fixed exchange rates led to capital controls that, ironically, put the banks into the business of direct lending to LDCs. After floating exchange rates and the oil crisis, the United States pumped up demand, kept its own oil prices low, and encouraged imports, virtually guaranteeing a second oil price surge. In the process the United States almost wrecked its currency. The resulting correction needed to be abrupt and, after a few false starts, Paul Volcker made it so.

The response to U.S. hard money policies after 1980 was the usual ambivalent one. Countries that had asked for it felt its reality too harsh. It emerged as a bizarre combination of tight money and heavy spending. This amounted to a bizarre combination in Keynesian terms—and most policy makers think Keynesian, especially in Europe and Japan—of flooring the accelerator while pressing hard on the brake pedal. The result was to squeeze all but the strongest borrowers out of the credit markets. The first of the government casualties surfaced in Washington, D.C., on August 13, 1982.

The Immediate Problem: Paying the Interest

While the LDC debt crisis is, as we have said, not a very subtle matter, the consistency with which analysts have avoided addressing it straightforwardly is both fascinating and revealing. Although it has been examined by everyone from Eliot Janeway, peering coolly over his half-circle reading glasses, to the frantic young São Paulo secretary so fearful of a social "explosion" that she and her unemployed husband left the city for the remote Brazilian countryside, some basic facts about the big five Latin debtors (Brazil, Mexico, Argentina, Venezuela, and Chile), together with Poland, Rumania, Yugoslavia, and the Philippines, are painfully clear. Of course they cannot repay their debts—any more than a family could pay off a new mortgage if asked to do so on a few day's notice. No one is really asking for or expecting repayment. The immediate problem—the debt crisis, if you like—is that, for the most part, these countries cannot pay the *interest* on their debts. Again, the analogy with a new homeowner is revealing. Mortgage contracts clearly stipulate a procedure whereby nonpayment of interest eventually results automatically in reversion to the bank of ownership of the house. Since banks seldom lend more than 80 percent of the purchase price in mortgage loans, the house of a laggard borrower can be sold, at a discount if necessary, to enable the bank to recoup its loan principle and any interest foregone during the year or so it might take to wrest ownership away from a deadbeat unable to pay interest and amortization (principle payback) charges. Mortgages are usually designed to repay interest and principal over twenty-five or thirty years, a generation's span. Nations live on forever, or so the anthems say. In any case, the big ones usually outlast a single lifetime and so amortization is not as crucial as it is with loans to mortal human beings.

But interest must be paid. The problem now being faced is what to do if it is not, since banks are in no position to attach significant assets belonging to debtor countries. Attempts during 1983 by small regional banks to grab an airplane here or there in order to collateralize their $5 million or so participation in a big Latin loan set off alarm bells at New York law offices and government agencies in Washington that quashed such initiatives like a sledgehammer slamming down on a fly. Since there is not anywhere nearly enough collateral to go around, the last thing the big banks want is a rush to grab airplanes or real estate belonging to Argentina or Brazil that would leave them very obviously very short of real collateral against their $200 billion or so exposure.

The considerable tension that is mounting around the inability of LDC borrowers to pay the interest charges on their hundreds of billions in debt arises from the fact that, in order to avoid having such payments declared in arrears by regulators, the banks are themselves lending the billions required to make timely interest payments. It is as if a local bank with about 10 percent of its assets—about twice its shareholder of equity—tied up in mortgages started lending destitute homeowners the money with which to pay their interest. The process presumes a sharp turnaround in fortunes of the borrowers since it results in a heavier exposure of the bank to borrowers who are—at least currently—clearly unable to service their debts. If the banks lend LDCs the funds to pay their interest charges, the $200 billion or so at risk in 1983 will double in about five years, given current interest rates of about 14 percent. The $700 billion LDC debt total would reach $1 trillion in just two years and nine months. The break comes when the banker—fearful of throwing good money after bad—loses his nerve and stops lending to borrowers who show no signs of reviving after huge transfusions of funds. Then, by Janeway's tidy definition, we have bankruptcy and default.

Will Lending Funds to Pay Interest Work?

For the optimists in the world of banking, the Mexican rescue package of December 1982 and the Brazilian rescue package of December 1983, which do no more than provide those countries with funds to make timely interest payments through the end of 1984, will carry borrower and lender to a brighter world where those countries, together with all the others that have been similarly rescued, will be able to carry—pay the interest on—debts some 15 percent larger than the ones they could not manage at the end of 1982. That happy state of affairs would require real growth well in excess of the real (after-inflation) interest burden of around 9 to 10 percent implied by such loans. At the end of 1983 the IMF, which could be forgiven a temptation to lean on the optimistic side, was forecasting for 1984 average real growth of 3.7 percent for non-oil developing countries compared to 1.6 percent in 1983 and the 0.9 percent realized in 1982. But among the troubled Latin group things looked worse. In Brazil 1983 real growth was running at minus 2–3 percent, with the economy strangled by austerity aimed at squeezing down consumption to free up funds to pay the loans. And investment-hampering, chaotic inflation was approaching 200 percent a year. Argentina was just emerging from its election, deciding how hard to press its military establishment for redress of past human rights' violations, and looking for a moratorium on its debts with its economy in a shambles since the disastrous Falklands War. Venezuela, with two-thirds of its $25.3 billion in debt due in 1983 and 1984 —in 1981–82 you could lend short to an oil exporter since the risk was (supposedly) low—was the new "sick man" of Latin America, with its 1983 oil revenues estimated at $14 billion, $5 billion below 1981's peak.[3] Even Mexico—everyone's candidate for a gold star in 1983—was still struggling with a faltering economy and problems of political stability after over a year of heavy economic sacrifice.

The requisite, sparkling, real growth required for heavy debtors in 1984 under the program of crisis aversion by lending interest was—everyone knew —a virtual impossibility. In the new marketplace where banks swap LDC debts in hopes of juggling exposure, *Business Week* reported in December 1983 that Brazilian debt was bringing 72 to 75 cents on the dollar; Mexico, 85 cents; Venezuela, 80 to 85 cents; and Argentina, 77 to 80 cents.[4] In the London secondary market Brazilian loans were selling for 60 cents on the dollar, according to the December 5 issue of *Forbes.*[5] Whatever the exact number, the Latin loans—not to mention the total of over $75 billion in loans to Poland, Yugoslavia, Rumania, and the Philippines—were not worth the 100 cents on the dollar at which they were being carried on the books of the big banks.

How Much Can Be Salvaged?

There is, in the light of the debt-crisis-induced down-valuation of a huge chunk of bank assets, a fundamental issue before the banks and the LDC borrowers together with the governments and international agencies like the IMF and the BIS. Should they try to keep the whole debt and its cargo vessel afloat with a part of it severely damaged, risking the possibility that the shattered section overburdened by cargo will drag the whole system under, or should they jettison cargo and sever damaged sections to ensure survival of the rest? The answer, of course, depends on how rough the seas are—a strong recovery would help to smooth the seas—as well as on extent and repairability of the damage. Relevant also are the possible proximity of repair vessels and the capacity of the pumps—rescheduling or rescue plans—to pump out water faster than it spills into the hull.

The point—on one side of which the debt vessel slips beneath the waves, while on the other it limps back to port for repairs that restore its seaworthiness—can be identified fairly accurately. The maximum viable ratio of Third World debts to the ability of Third World governments to pay them clearly —in the light of the current "crisis"—is somewhat below current levels. That fact gives us a clear idea of conditions that would have to be met to get the debt vessel back to full seaworthiness without drastic surgery. The ability of governments to pay debts relates to government revenues, which in turn depend on the average tax rate multiplied by GNP, the tax base. We need to identify the conditions under which the ratio of debt to governments' ability to pay stays constant—a minimum condition for long-run survival given demonstrable inability of LDCs to service current debt levels without new money. The ratio of debt burden to ability to pay stays constant if the growth rate of the debt burden equals the growth rate of ability to pay. If borrowers are borrowing to pay interest, the debt burden grows at the interest rate—the sum of the real interest rate and expected inflation. Ability to pay grows at a rate equal to the growth of GNP—the sum of inflation and real output growth—plus the rate of increase in the average tax rate, which of course is limited by willingness of the electorate to take on a rising tax burden.

Summarizing, a stable ratio of debt burden to ability to pay, the minimum viable condition for trying to get out of the debt crisis hole, requires:

(real interest rate) + (expected inflation) =
(growth of tax rate) + (inflation) + (growth of output)

On average, expected inflation is roughly equal to inflation so those terms cancel each other, leaving:

(real interest rate) = (growth of tax rate) + (growth of output)

This simple expression captures most aspects of many popular discussions of the debt crisis. If real interest rates are higher than the sum of terms opposite, the ratio of debt burden to ability to pay will rise, making collapse very likely. Therefore, high real interest rates are decried as a cause of the debt crisis and identified as needing to fall to help get us out of it. If tax rates hold steady, then the only antidote to high real interest rates is rapid growth of output, which unfortunately becomes less likely at high real interest rates. Some analysts use growth of exports in place of output growth since Third World debts are to foreigners and foreign exchange must be earned to pay them, barring further net infusions of foreign exchange from capital inflows. If all inflows are just flowing out again as interest payments, then exports of goods and services provide the only foreign exchange. Over the short term as well as over the long term, barring major export promotion drives supplemented with heavy investment in export industries, the share of exports in total output remains about constant so that export growth is roughly equal to overall output growth. Obviously it is tempting to hope for an export surge to help out, but for LDCs a strong world economic recovery is the only basis for a sustained rise in exports. In the open LDC economies, export growth is really just a proxy for the overall economic growth needed to increase debt servicing capacity.

We have already seen that real interest rates facing the Third World in 1983 were about 9 percent, while 1984 projected growth rates averaged only about 3.7 percent, with considerably lower—even negative—rates likely for biggest Latin debtors like Brazil, Argentina, Venezuela, and Mexico. These conditions fall far short of the minimum viable case where output growth would have to equal or exceed real interest rates provided that tax rates are stable. Of course that latter provision can be relaxed, but such action carries with it heavy risks of the type that show up on newsreels of angry Brazilian workers looting supermarkets in São Paulo.

The fudge factor usable when real interest rates exceed growth rates is higher taxes, usually called austerity. If an economy is not generating the revenues a government needs to pay interest charges at current tax rates, the only available expedient besides default is to raise tax rates. In LDCs this is usually done with an inflation tax, with the government just printing the money it spends instead of borrowing it or collecting taxes directly. The "take" of such an inflation tax comes from the depreciation of government liabilities to its people—the unindexed money and bonds they hold—that is caused by inflation. We have already described a milder version of the tax collected from Americans since World War II. Three-digit inflation rates like those in Brazil set the tax rate much higher. The effects of such astronomical rates can be devastating. Many wages and prices are indexed—tied to inflation—but for those that are not, like the incomes of the poor, already limited purchasing power can be virtually wiped out. Further, inflation begets more inflation as people rush to "buy now" before prices rise again, and it is easy for triple-digit inflation rates to surge to the incredible, almost infinite rates seen during the German Weimar government's hyperinflation of the 1920s. At that point the inflation tax becomes totally fruitless—the money literally is not worth the paper it is printed on—and the purpose—raising taxes to stabilize the ratio of debts to means of payment—is long since forgotten and, of course, unachieved as well.

Looking ahead from 1983—as well as looking ahead from 1984—it seemed very unlikely that the minimum condition for "muddling through" the debt crisis—a phrase favored by bankers pressed for opinions on how they thought the crisis might be resolved—could be met in the foreseeable future. Under the most optimistic assumptions, the gap between the Third World's real interest rates and real growth yawned wide at somewhere between 5 and 6 percentage points—not 5 or 6 percent. Six percent meant that the real debt burden of already overburdened Third World countries would rise by 50 percent by the end of the decade. An attempt to close such a broad gap with heavier taxes—directly or indirectly with inflation—carried with it a high probability of a total collapse of government credit in the Third World. The social chaos that accompanies hyperinflation and unbearable tax burdens would result in political upheaval and, thereafter, new governments that almost by definition would be committed to repudiation of what would be termed "illegitimate" debts of an "oppressive" regime that had pressed too hard to wring the last peso, cruzeiro, or bolivar out of its citizens and, in doing so, made those currencies worthless.

During 1983 one frequently heard observers who espoused the "muddling through" view about the debt crisis go on to explain that such a result was likely because it was neither in the best interest of the banks or the borrowing countries to do otherwise. Such a view is naive at best and dangerous at worst. War is in no one's best interest but it happens. Just because two explorers lost in the jungle decide to work together to find their way out, there is no guarantee that they will succeed. Other powerful forces like weather, wild animals, and simple luck or a lack of it affect their fate, just as the rioting poor, vagaries of nature, and markets with their effects on commodity prices and plain old luck will have important effects on the outcome of the debt crisis. The simple fact is that no one knows how the crisis will be resolved, but analysis of the basic data on real interest costs and output growth together with the potential to raise more tax revenue from the already frustrated population of the Third World strongly suggest that an attempt to "muddle through" carries with it a significant risk of total collapse. Prudent men should take note, anticipate such a problem, and take steps to avoid it.

The required steps are clear and painful, especially for the big banks. Another way of stating the problem is to acknowledge that secondary markets for Latin debt placed a value on it at year-end 1983 of between 60 and 85 cents on the dollar—say an average of 75 cents on the dollar, allowing for a heavy weight given to serious problem cases of Brazil, Argentina, and Venezuela. Good data are available on exposure of the ten largest U.S. banks to the five largest Latin borrowers. From this we can calculate roughly the magnitude of current losses that the market value of debts of these countries implies for the big-10 banks and how such losses might be absorbed into their earnings.

Impact of Losses on the Banks

Like any household or business, a bank that sustains a loss in the value of its assets must take the funds to restore assets out of income. For significant losses, such as collapse of a volatile stock for a household or a 25 percent write-down on Third World loans—for the big-10 U.S. banks' loans to the big-5 Latin borrowers as of September 30, 1983 totaled $44.2 billion—absorption of asset losses from income can be painful since assets are usually a multiple of annual income.[6] As of September 30, 1983, equity of the big-10 U.S. banks, their assets reserved for shareholders over and above liabilities to depositors and other creditors stood at $30.93 billion, while total assets as of end-1982 stood at about $625 billion. The big banks are highly leveraged with equity only about 5 percent of total liabilities, and therefore they have little leeway to cut equity further. *Euromoney* reported that in 1982 U.S. banks earned an average of 12.7 percent on equity while the big-10 earned 12.3 percent on equity.[7] If we are optimistic and assume that they earn 14 percent on equity in 1983, big-10 earnings will be about $4.33 billion ($30.93 billion times 0.14).

What about big-10 losses on their $44.2 billion in loans to the big-5 Latins, evaluating them at 75 cents on the dollar? These losses come to a staggering $11 billion, or two and one-half years' earnings—not including possible losses on other loans to the Philippines, Poland, Rumania, and Yugoslavia, to mention only the largest of other problem borrowers. It is calculations such as these that cause bankers and bank stock analysts to shudder. Even if the $11 billion were absorbed into earnings over ten years, the first-year write off of $1.1 billion represents more than 25 percent of big-10 bank earnings. And if, in fact, 1983 big-10 earnings equal the average of a 12.7 percent return on equity—still better than last year's 12.3 percent—their $3.93 billion in earnings would be even more vulnerable with $1.1 billion write-offs absorbing 28 percent of profits and a total of $11 billion in losses eating up two years and nine months' worth of earnings.

These grim facts were not ignored by the stock market. Historically, shares of the big banks have traded at about 80 percent of the price/earnings (P/E) multiple of the overall Standard and Poor's (S&P) 500 stock index. The November 1983 S&P multiple of about 11 would mean that the big banks' P/E ought to be about 8.8. As of November 1983, Citibank's P/E was about 4.8; Bank of Boston's, 4.6; First Chicago's 5.3; Chase Manhattan's, 3.7. Even relatively conservative Morgan's was only 5.5.[8] (By June 1984, average money center banks' P/E's had dropped to 4.5 with Manufacturers Hanover's down to 3.5.) A P/E of 5.8 would be two-thirds of the "normal" bank P/E of 8.8, consistent with a loss of one-third of normal bank earnings. It appears that the stock market was taking very seriously the possibility that banks would have to absorb LDC loan losses equal at least to between one-quarter and one-third of annual earnings for the foreseeable future.

The only degree of freedom the banks and their regulators have once the decision is made to write-down assets is the speed at which such write-downs are absorbed into earnings. Given an interest rate of 10 percent, the present value of an $11 billion write-off spread out over ten years of payments at $1.1 billion per year is about $6.75 billion; five years of $2.2 billion payments are worth $8.33 billion in present-value terms, while twenty years of $0.55 billion payments carry a present value of $4.68 billion. This arises from the fact that, given an opportunity to earn 10 percent interest, a dollar of debt to be paid next year has a present value of 90.91 cents, since if you put aside 90.91 cents today, in a year's time it is worth a dollar (90.91 times 1.1 = 1.00). Obviously, if the banks could convince the regulators to let them write-off the $11 billion over one hundred years at $110 million per year, they could absorb—albeit with some pain—the $1.099 billion present value of the loss in one year. But the more slowly the loans are written off, the more distorted are current "earnings" figures as a measure of true profits. Ten years appear to be about as far out as regulators will contemplate going. That still leaves losses conservatively estimated to have a present value worth nearly two years' earnings for the big banks, especially when actual and prospective nonperforming loans outside of Latin America are included in the prospective loss figures.

Responses by individual banks to prospective loan losses have varied. Available data[9] on loan-loss reserves as of September 30, 1983, showed different degrees of concern at the biggest banks measured by the share of loss reserves in total assets (in parentheses): Citibank, $728 million (0.82 percent of assets versus 0.76 percent on 9/30/82); Chase, $552 million (1.02 percent of assets versus 1.01 percent on 9/30/82); Bank of America, $400 million (1.21 percent of assets versus 0.87 percent on 9/30/82); Manufacturers Hanover, $382 million (0.83 percent of assets versus 0.82 percent on 9/30/82); Morgan, $446 million (1.39 percent of assets versus 1.02 percent 9/30/82). While these loan-loss reserves total $2.5 billion, all are not available to cover losses on LDC debts. In fact, most of the reserves were already in place by September 30, 1982, indicating that they were viewed as part of normal, prudential reserves. What is remarkable is that of the five largest and most exposed U.S. banks only two, Bank of America and Morgan, added significantly to loan-loss reserves between September of 1982 and September of 1983. Add to this the fact that 1983 reported earnings are overstated by virtue of the fact that the big banks were counting as earnings hundreds of millions in interest due from Latin debtors that were not actually received, and it is little wonder that by November 1983 the P/E for many big banks was not over one-half its historical average.

Possible Outcomes

The unpleasant realities about the big banks' exposure had, by the end-1983 breathing spell after the U.S. Congress's passage of IMF quota legislation and an apparently successful scramble to scrape together funds to enable Brazil to pay interest on its debts through the end of 1984, sunk in with a resounding thud. The debt crisis—notwithstanding the *Wall Street Journal*'s April 20, 1983 editorial pronouncement that "the international debt crisis . . . is for all practical purposes, over"—was real, and not over.[10] Three possible ways out were envisioned. The first involved the process already underway in 1982–83 of rescheduling debts to longer maturities and spreading out repayment of interest—possibly at lower rates—and principle over a longer period of time in the apparent hope that Brazil, Argentina, and other big debtors would somehow better be able to pay later the debt-service charges they could not pay in 1983. The second solution, one that by the end of 1983 had emerged as only a partial remedy, was a strong and sustained world economic recovery. While faster growth would obviously help, given the gap between real interest rates and optimistic projections of growth—together with Morgan Guaranty's projections of billions in additional funds needed to service LDC debts even with a strong recovery—the growth "solution" to the debt problem was a little like visions of a spring blizzard dancing in the head of the owner of a warehouse full of tire chains after a warm winter.

The most widely touted long-term solution to the debt crisis as 1983 turned into 1984, with its strong Orwellian overtones, was the "finance company" approach. In exchange for getting out from under their LDC loans, the banks would turn them over, at a discount, to some government agency that would then issue long-term bonds to replace troublesome paper on the banks' asset rosters. The agency, presumably endowed with great powers, would then deal with the debtors, possibly tying a scaling down of their foreign debts to a scaling down of government domestic debt. Lower fiscal deficits and less inflationary monetization of those deficits, with resultant restoration of monetary order, would be the price for lightening the Third World's foreign debt burden. Various elements of such a constructive *quid pro quo* proposal have been put forward by a number of analysts of the debt problem, including MIT economics professor Rudiger Dornbusch, British peer Lord Lever, Princeton economist Peter Kenen, and architect of the New York City financial restructuring, Felix Rohatyn, to mention just a few.

Who Pays?

All of the proposals, either directly or indirectly, acknowledge, first, that the banks' LDC loans are not worth 100 cents on the dollar and, second, that some consolidation tied to constructive adjustment by both debtor and creditor is desirable. All leave unanswered the questions of at what levels below 100 cents on the dollar will the varied loans to varied borrowers be valued and what "agency" will serve as consolidator, issue long-term bonds to be held in place of LDC debts, and—most important—enforce the *quid pro quo* conditions that would have to be accepted by debtor countries. Finally, there is the biggest question of all: just who will absorb the losses, which could run as high as $70 billion at an average of 90 cents to the dollar on the total of $700 billion in Third World debt to banks, governments, and international agencies?

Perhaps even more dicey—particularly in the shorter run—is the question of whether the commercial banks will be forced to swallow the $30 to $40 billion in losses on their roughly $300 billion share of that debt. In principle, there is no reason why they should not, as Milton Friedman, the *Wall Street Journal,* and congressional hard-nosed conservatives have frequently reminded us since 1982. For their part, the bankers, with frequent references to the "responsibilities" they undertook—with strong government encouragement, they add in unconvincingly reluctant tones—express different feelings. Governments and international organizations like the IMF and the BIS are not saying. They are too busy trying to avoid a premature collapse that would eradicate the luxury of making any nice orderly judgments about "who pays."

Recalling our earlier discussion of crisis phases, it is to be hoped that the banks and their governments will move quickly from stages of denial and panic to stages of confrontation and absorption. This will require an open acknowledgment by the bankers that they are not going to come out of the debt crisis whole. Those who resist this painful step need only look at market evaluation of their stock to realize that reality has overtaken them. It may be that the uncertainty associated with the wide range of possible outcomes, including a possible financial collapse, are raising the banks' costs of capital—especially costs of less involved banks or those like Bank of America or Morgan's, which have reacted by adding significantly to loan-loss reserves since 1982—by more than would be the case after all had "swallowed their medicine" and got on with business.

If the "finance company" approach is adopted, an agency, probably the IMF —the United States will just have to leave aside its political objections here— should start negotiations that are aimed at parceling out losses while tying constructive programs, such as reduction of the Third World's fiscal deficits and less inflationary financing, to alleviation of debt burdens. Losses to be borne could be estimated from the price—below par—at which some agency like the IMF could market longer term bonds backed by claims on all debtor countries, with some help coming from the diversification element implicit in buying a claim on a package of countries. Some allocation of losses that individual participants see as "unfair" would very likely result, but such losses would fall far short of gross losses that could result from the collapse that will come if the debt problem goes unaddressed.

Like many problems that essentially need to be faced, it is far better to address the debt crisis suboptimally than not to address it at all. If fears over "who pays how much?" or "what agency?" paralyze action, eventual losses will be far greater than those resulting from a possible "unfair" parceling out of some bank's or some government's share of losses. Going even further, the late Sir Harry Johnson often said, "Sometimes it's better to make the wrong decision than to make no decision at all"; in the case of the debt crisis, with a nondecision significantly increasing the possibility of a financial collapse, Sir Harry's dictum is probably valid. At the very least, legitimate *ex ante* concerns about the fine points of a program to address the debt problem and to parcel out the losses, which have already occurred and which will not be reversed, should not be allowed to prevent bankers, the governments, the LDCs, and the IMF from squarely addressing the very real debt problem.

The Larger "Debt Problem"

Before finishing our story, it is important to place the debt crisis of the eighties into a broader context. The early 1980s has seen the confrontation phase of a general crisis resulting from rapid expansion of government debt of both developed and developing countries. The increase in debt of governments of developed countries has not yet reached a widely recognized crisis stage, although intense 1983–84 U.S. debate over deficits—additions to U.S. government debt—suggests that some are anticipating a crisis. The LDC debt crisis is just the leading edge of a broader debt crisis that is grinding toward us like a glacier creeping down a valley threatening to bury a stand of trees in its way.

Everywhere the explosion of government debt mirrors the legislation, largely during the 1960s, of economic hopes for the future into virtual economic requirements. Unfortunately, those requirements, largely for sustained economic growth of 4 to 6 percent annually, have gone unmet in industrial countries since 1973 and in developing countries since 1980. The postwar industrial world came of age in the early 1960s. In the United States of 1961, the forty-three-year-old new president, John F. Kennedy, ascending toward the peak of his adult powers, mirrored the progress of a nation to which many things, including eradication of want, travel to the moon, and continued economic growth, all seemed possible. Kennedy's successor, Lyndon Baines Johnson—a powerful executive with a genius for moving legislation born of years in Congress—got the Kennedy program—dubbed the Great Society— passed into law. Unfortunately, he also tacked on the huge requirements of financing a tragic Vietnam War that was to drag on for a decade until the sorry April 1975 spectacle of Americans being plucked by helicopters off the roof of a beseiged U.S. embassy in Saigon. That war, together with the 1973–74 oil shock—made possible by inflationary finance employed by three U.S. governments to pay for both guns and butter—invalidated the rosy economic assumptions that had convinced the senators and representatives on Capitol Hill to expand lavishly social programs while always expecting an imminent end to the heavy demands of a war that everyone wanted to have done with.

The U.S. expansion of social programs mirrored one that was underway in Europe and Japan. When the big industrial countries were slapped with a $70 billion oil surcharge early in 1974, the panic stage of a very slowly unfolding global economic crisis set in, reinforced by the gloomy portents in the Club of Rome's widely touted *Limits to Growth,* which had appeared in 1972. Four years of aggressive demand expansion, from 1974 to 1979, especially in the United States, were a drawn-out denial phase of the crisis, akin to flogging a tired horse in the vain hope that somehow beating the exhausted beast hard enough would restore its energy. The most intense flogging was administered by a righteous engineer, Jimmy Carter, and his team of economic dreamers who chanted in unison that it was all for our own good.

Half a decade of denial brought a second oil crisis—a second crisis being the inevitable result of protracted denial of a first—and with it a painful period of confrontation, especially after 1980. The Third World debt crisis that surfaced in August of 1982 with its own chronology of panic, denial, confrontation, and eventual absorption was superimposed, like a swell atop a tidal wave, on the confrontation phase of a more slowly unfolding, even larger crisis. The realization dawned that governments everywhere had promised more than they could ever deliver without creating a mountain of debt and thereby leaving themselves with the painful choice of printing money to collect the insidious inflation tax to pay for it, cutting back on promises, or putting through sharp increases in direct taxes. The rule whereby sustainable debt burdens require real income and tax rates to rise as fast as real interest rates applies for both developed and developing countries. It also applies—as Americans will discover—to countries that add two dollars in tax cuts and military spending for every one dollar of reductions in government spending on social programs.

The confrontation phase and perhaps the start of the painful absorption phases of the world debt crisis and its most acute aspect, the Third World debt crisis, descended on us as 1983 turned into 1984. Somewhat ironically, this crucial stage of a compound debt crisis came in the midst of a heartening economic recovery, especially in the United States, where unexpectedly sharp drops in unemployment and inflation lowered the "discomfort index," which records the combined effect of those twin maladies. But looking ahead, the U.S. government saw its debts piling up at rates of about $200 billion per year, with annual interest payments on the debt already accumulated approaching $100 billion and projected to rise to over $140 billion in 1986. Financing such deficits together with other borrowing needs required, and would continue to require, tremendous flows of capital to the United States from the rest of the world, $40 billion in 1980, $75 billion in 1981, and $85 billion in 1982.[11] These huge sums were pulled in by the high real interest rates being paid by American borrowers lining up behind the U.S. Treasury, which was shouldering its way to the front of the line to gobble up almost $20 billion a month off the top. The result would be extra pressure on already-strained credit markets abroad in countries whose governments had their own deficits to finance.

In spite of the large deficits in the United States and elsewhere in the developed world, the debt crisis facing the LDCs at the end of 1983 was far more acute. Borrowing by the big-5 Latin borrowers had risen by an average of almost 25 percent annually for the years from 1973 to 1982. Even projections of $200 billion U.S. deficits through 1987 would imply a growth rate of about 14 percent for U.S. debt over the half decade after 1982—well below the growth rate of Latin foreign debts.

But there were still some troublesome aspects of the big increases in U.S. debt. Just as the LDC debt crisis, born of 20-percent-plus annual increases in LDC debts to foreigners, was unfolding, U.S. borrowing from abroad was rising at an annual rate of 20 percent between the end of 1980 and the end of 1983. True, the ability of a widely diversified U.S. economy to service debts largely denominated in its own currency in a world where the U.S. central bank can do a great deal to control interest rates is far greater than that of the

For Whom the Bell Tolls

LDCs. But as we face up to the consequences of a Third World borrowing binge—as we must to avoid facing a wave of financial panics akin to those in 1929 and the early thirties—it is well to remember words written over three centuries ago, in 1624, by John Donne:

> No man is an island, entire of itself; every man is a piece of the continent, a part of the main; if a clod be washed away by the sea, Europe is the less, as well as if a promontory were, as well as if a manor of thy friends or of thine own were; any man's death diminishes me because I am involved in mankind; and therefore never send to know for whom the bell tolls; it tolls for thee.[12]

Notes

Introduction

1. Anthony Sampson, *The Money Lenders* (Middlesex and New York: Penguin, 1981), pp. 18–19.

Chapter 1

1. "U.S. and Mexico: Major Rift Emerges," *New York Times,* 14 August 1982, p. 2.
2. Ibid.
3. Wolf Von Eckardt, "A Brilliant Example of Inner Space," *Washington Post,* 18 August 1973, pp. B1–B3.
4. See "The Dynamite Issue," *Newsweek,* 30 May 1983, pp. 22–24. These figures are estimates later largely confirmed by 1983 Securities and Exchange Commission–mandated release of banks' loan exposure by country. Of the ten largest U.S. banks, only the largest, Citibank, declined to disclose data as of September 30, 1983, prior to the period of mandatory reporting that began at the end of 1983. See also "Problem Latin Loans Hit 7% Level," *American Banker,* 5 December 1983, pp. 1–3.
5. Les Aspin, "It's Not Just a Bail-out for Banks," *Washington Post,* 20 January 1983, p. A19.
6. See statement by Paul A. Volcker before the Committee on Banking, Finance and Urban Affairs, 2 February 1983.
7. "The Shunning of the Sovereign Borrower," *Euromoney,* May 1982, p. 41.
8. See statement by Volcker before the Committee on Banking, Finance and Urban Affairs, 2 February 1983, p. 8.
9. See Milton Friedman and Anna J. Schwartz, *A Monetary History of the United States* (Princeton, N.J.: Princeton University Press, 1963). The chapter entitled "Great Crash" is especially revealing.
10. Henry James, *The American* (New York: New American Library, 1980), p. 28.
11. Friedman and Schwartz, *Monetary History,* pp. 310–11.
12. Ibid., pp. 327–28.
13. Raymond de Roover, *The Medici Bank* (New York: New York University Press, 1948).
14. Data are from the International Monetary Fund, *Annual Report* (Washington, D.C.: IMF, 1983), chap. 1.
15. Ibid.
16. Ibid.
17. Statement by Paul A. Volcker before the House Committee on Banking, Finance and Urban Affairs, 2 February 1983.
18. J. M. Keynes, *The Economic Consequences of the Peace* (New York: Harcourt Brace, 1920), pp. 280–81.
19. Ibid., p. 281.

Notes

Chapter 2

1. David Hume, "Political Discourses, 1752," in Eugene Rotwein, ed., *David Hume: Writings on Economics* (London: Nelson, 1955).

2. Walter Bagehot, *Lombard Street* (London: Kegan Paul, 1904).

3. Charles Dickens, *A Christmas Carol* (New York: Washington Square Press, 1963), p. 60.

4. "Frank & Ernest," by Thaves, MEA, Inc., syndicated October 22, 1983.

5. Data on sources and uses of international capital flows are from Derek Aldercroft, *From Versailles to Wall Street* (London: Allen Lane, 1977), chap. 10.

6. See Ralph Reisner, "Default by Foreign Sovereign Debtors: An Introductory Perspective," *University of Illinois Law Review,* no. 1 (1982): 19.

7. Ibid., p. 20.

8. Ibid.

9. Aldercroft, *From Versailles to Wall Street,* chap. 10.

10. Joseph S. Davis, *The World Between the Wars—1919–39: An Economist's View* (Baltimore: Johns Hopkins University Press, 1975), pp. 148–49.

11. Anthony Sampson, *The Money Lenders* (Middlesex and New York: Penguin, 1981), pp. 14–15.

12. F. Scott Fitzgerald, *The Great Gatsby* (New York: Charles Scribner's Sons, 1925), p. 182.

13. H. E. Peters, *The Foreign Debt of the Argentine Republic* (cited in Aldercroft, Chap. 10, p. 250 1934).

14. International Monetary Fund, *Annual Report* (Washington, D.C.: IMF, 1983), chap. 1.

15. John Kenneth Galbraith, *The Affluent Society* (Boston: Houghton Mifflin, 1958).

16. D. H. Meadows, et al., *The Limits to Growth* (New York: Universe Books, 1972).

Chapter 3

1. J. M. Keynes, *The Economic Consequences of the Peace* (New York: Harcourt Brace, 1920), p. 25.

2. Alfred Marshall, *Principles of Economics* (London: Macmillan, 1890).

3. Robert Torrens, *Essay on the Production of Wealth* (London: N.p., 1821).

4. J. M. Keynes, *The General Theory of Employment, Interest and Money* (London: Macmillan, 1936), p. 383.

5. Keynes, *Economic Consequences,* p. 20.

6. See A.J.P. Taylor, *English History 1914–1945* (Oxford: Clarendon Press, 1965).

7. These and subsequent data on economic performance are from the International Monetary Fund, *Annual Report* (Washington D.C.: IMF, 1982, 1983), chap. 1.

8. Keynes, *Economic Consequences,* p. 39.

9. Ibid., p. 41.

10. Henry James, *The American* (New York: New American Library, 1980), pp. 87–88.

11. D. H. Meadows, et al., *The Limits to Growth* (New York: Universe Books, 1972), p. 23.

12. Data are from the International Monetary Fund, *Annual Report* (Washington, D.C.: IMF, 1983), chap. 1.

Chapter 4

1. See H. W. Singer, "The Distribution of Gains Between Investing and Borrowing Countries," *American Economic Review* 40 (May 1950): 473ff., and R. Prebisch, "Commercial Policy in the Underdeveloped Countries," *American Economic Review* 49 (May 1959): 251ff.

Notes

2. See United Nations, *International Financial Statistics,* 1966–67 supp. (New York, 1965), pp. xiv–xv.

3. For a general discussion of developments in Latin America since the early 1960s, see David Collier, ed., *The New Authoritarianism in Latin America* (Princeton, N.J.: Princeton University Press, 1979).

4. Ibid.

5. Robert R. Kaufman, "Industrial Change and Authoritarian Rule in Latin America," in Collier, ed., *New Authoritarianism,* pp. 165–254.

6. Figures cited in Collier, ed., *New Authoritarianism,* p. 236.

7. Earl Cowley, Speech to the British House of Lords, *Review of the River Plate,* 29 February 1972, p. 250.

8. These and other figures on Brazil are drawn from: Collier, ed., *New Authoritarianism,* pp. 165–253; International Monetary Fund, *Annual Report* (Washington, D.C.: IMF, 1982); and William R. Cline, *International Debt and the Stability of the World Economy* (Washington, D.C.: Institute for International Economics, 1983).

9. Data are from the International Monetary Fund, *International Financial Statistics* (Washington, D.C.: IMF, 1983), chap. 1.

10. See Cline, *International Debt.*

11. See Singer, "Distribution of Gains," and Prebisch, "Commercial Policy."

12. "Country Risk League Tables," *Euromoney,* October 1979, p. 130.

13. Cline, *International Debt,* pp. 130–31.

14. Ibid.

15. *Euromoney,* October 1981, p. 159.

16. Council of Economic Advisors, *Economic Report of the President* (Washington, D.C.: U.S. Government Printing Office, February 1982).

17. Data are from *Monetary Trends* (St. Louis: Federal Reserve Bank of St. Louis, March 1983, June 1983).

18. "Brazil: A Survey," *Economist,* 12 March 1983, pp. 1–18.

19. Pedro-Pablo Kuczynski, "Latin American Debt: Act Two," *Foreign Affairs* Fall 1983, pp. 17–38.

Chapter 5

1. See Federal Reserve Board, *Federal Reserve Bulletin* (Washington, D.C.: U.S. Government Printing Office, various issues, 1960–61).

2. Robert Triffin, *Gold and the Dollar Crisis* (New Haven: Yale University Press, 1960), p. ix.

3. Taken from published text of Kennedy speech, U.S. Congress, Senate, *The Freedom of Communication: Part I Kennedy Speeches,* S. Rept. 994, 87th Cong., 1st sess. (Washington, D.C.: Government Printing Office, 1961), pp. 557–566.

4. Report on reaction by London papers to gold bubble, *New York Times,* 22 October 1960, p. 35.

5. Ibid.

6. Reported in *Wall Street Journal,* 24 October 1960, p. 12.

7. Ibid.

8. Council of Economic Advisors, *Economic Report of the President* (Washington, D.C.: U.S. Government Printing Office, February 1968).

9. *Economic Report of the President,* February 1984.

10. These and any figures on growth of money, gold, or foreign dollar holdings can be found in the *Federal Reserve Bulletin* appearing several months after dates being reported. Other statistics on growth of the U.S. economy or inflation rates are available in either the *Federal Reserve Bulletin* or the *Economic Report of the President.*

11. Leonard Silk, "Money Crisis Impact," *New York Times,* 12 May 1971, p. 55.

12. Richard Jannsen, "Cause of the Crisis," *Wall Street Journal,* 7 May 1971, p. 1.

13. Ibid.

14. Ibid.

15. John M. Lee, "Villain of the Monetary Crisis: Mysterious Eurodollar Market," *New York Times,* 10 May 1971, pp. 1, 52.

16. "Another Money Crisis," *Wall Street Journal,* 6 May 1971, p. 14.

17. "Commodities," *Wall Street Journal,* 6 May 1971, p. 22.

18. "Gold Outglitters the West German Mark," *Wall Street Journal,* 13 May 1971, p. 13.

19. *Wall Street Journal,* 10 May 1971, p. 1.

20. Quoted in Martin Mayer, *The Fate of the Dollar* (New York: Times Books, 1980), p. 179.

21. Otmar Emminger, in Clyde H. Farnsworth, "The Other Side of the Dollar: Europeans Describe View" *New York Times,* 26 May 1971, p. 55.

22. *Economist,* 25 December 1971, p. 10.

23. Ibid.

24. See footnote 10.

25. See footnote 10.

Chapter 6

1. Council of Economic Advisors, *International Economic Report of the President* (Washington, D.C.: U.S. Government Printing Office, 1977), pp. 175–76.

2. International Monetary Fund, *Annual Report* (Washington, D.C.: IMF, 1983), chap. 1.

3. Ibid.

4. These and other figures on economic performance of industrial countries are taken from the International Monetary Fund, *Annual Report* (Washington, D.C.: IMF, 1982, 1983).

5. Ibid.

6. R. E. Lucas, Jr., "Some International Evidence on Output-Inflation Trade-Offs," *American Economic Review,* June 1973, pp. 326–34.

7. J. M. Keynes, *The General Theory of Employment, Interest, and Money* (London: Macmillan, 1936).

8. IMF, *Annual Report.*

9. "Summit in Slump," *Economist,* 15 November 1975, pp. 81–84.

10. Ibid., p. 84.

11. Khodadad Farmanfarmaian, et al., "How Can the World Afford OPEC Oil?" *Foreign Affairs* 53 (January 1975): 201–11.

12. Edward R. Fried and Charles L. Schultze, eds., *Higher Oil Prices and the World Economy* (Washington, D.C.: Brookings Institution, 1975).

13. Edwin L. Dale, Jr., "What Happened to All the Oil Billions?" *New York Times,* 7 December 1975, p. E2.

14. Arnaud de Borchgrave, "Interview with Treasury Secretary Designate Michael Blumenthal," *Newsweek,* 27 December 1976, pp. 15–16.

15. IMF, *Annual Report.*

16. Ibid.

17. Ibid.

Chapter 7

1. Data are drawn from the Council of Economic Advisors, *Economic Report of the President* (Washington, D.C.: U.S. Government Printing Office, February 1983), and the International Monetary Fund, *International Financial Statistics* (Washington, D.C.: IMF, various years).

2. Data are drawn from the Department of Commerce, *Survey of Current Business* (Washington, D.C.: U.S. Government Printing Office, various issues).

Notes

3. Ibid.

4. Data are from the International Monetary Fund, *Annual Report* (Washington, D.C.: IMF, 1982), p. 36.

5. From Solomon Brothers, as cited in Anthony Sampson, *The Money Lenders* (London: Penguin, 1981), p. 178.

6. Ibid., p. 180.

7. "Why Citibank's Walt Wriston Is Looking Forward to the 1980s," *Euromoney,* July 1978, p. 84.

8. Ibid., p. 93.

9. Data on exposures in Latin America are from the London *Economist,* Financial Report, vol. 8, no. 186 (15 September 1983). Numbers vary somewhat depending on source and reporting data, but none alter the overall impression of very heavy exposure.

10. John Gunther, *Inside Europe* (New York: Harper, 1936) p. 256.

11. For an excellent and remarkably prescient—yet disquieting—discussion of legal precedents for government defaults on foreign borrowing, see "Symposium on Default by Foreign Government Debtors," *University of Illinois Law Review,* no. 1 (January 1982).

12. Interview with Anthony Sampson, cited in Sampson, *The Money Lenders,* p. 159.

13. Sampson, *The Money Lenders,* p. 181.

14. Ibid.

15. Ibid., p. 101.

16. S. C. Gwynne, "Adventures in the Loan Trade," *Harper's,* September 1983, p. 23.

17. Cited in "Philippines Debt Talks Beginning," *American Banker,* 18 November 1983, p. 2.

18. "Champion of the CITI," *Euromoney,* October 1983, p. 295.

19. Walter B. Wriston, "Banking Against Disaster," *New York Times,* 14 September 1982, p. A. 30.

20. Leonard Silk, "Less Jittery Financiers" *New York Times,* 15 September 1982, p. D 2.

21. "Financial Report," *Economist,* 15 September 1983, p. 1.

22. Ibid.

23. "Citibank's Walt Wriston," p. 92.

24. "Champion of the CITI," p. 296.

Chapter 9

1. William Manchester, *The Glory and the Dream* (New York: Bantam, 1975), p. 124.

2. Ibid., p. 125.

3. Roy F. Harrod, *The Life of John Maynard Keynes* (London: Macmillan, 1951), p. 445.

4. Fred Hirsch, *Money International* (London: Penguin, 1967), chap. 12.

5. Ibid., p. 238.

6. Royal Economic Society, *Collected Writings of J. M. Keynes,* vol. 25 (London: Macmillan, 1980), p. 195.

7. Bray Hammond, *Banks and Politics in America* (Princeton, N.J.: Princeton University Press, 1957), p. 359.

8. See Charles Coombs, *The Arena of International Finance* (New York: John Wiley, 1976).

9. Edward Jay Epstein, "Ruling the World of Money," *Harper's,* November 1983, p. 48.

10. Council of Economic Advisors, *Economic Report of the President* (Washington, D.C.: U.S. Government Printing Office, February 1984).

11. Robert Solomon, *The International Monetary System, 1945–1976* (New York: Harper & Row, 1977), pp. 86–95.

12. Data are from the International Monetary Fund, *International Financial Statistics* (Washington, D.C.: IMF, 1971).

13. Cited in Solomon, *International Monetary System*, p. 162. Originally from *New York Times*, 10 May 1969, p. 1.

14. Solomon, *International Monetary System*, p. 157.

15. See "Exchange Uncertainty," *Wall Street Journal*, 22 December 1976.

16. As reported by Hirsch, *Money International*, p. 243.

17. Ibid.

18. Ibid. ("There was never any gentlemen's agreement of Basel, but at Basel there are only gentlemen.")

19. Epstein, "Ruling the World of Money," p. 44.

20. Cited in ibid., p. 45.

21. Arlene Wilson, "The Stability of the International Banking System" (Congressional Research Service, Issue Brief IB82107, Washington, D.C., 19 July 1983).

22. For a full discussion of the Basel Concordat see Wilson, "International Banking System," pp. 7–8.

23. See Phillip Cagan, "The New Monetary Policy and Inflation," in *Contemporary Economic Problems* (Washington, D.C.: American Enterprise Institute, 1980), pp. 9–38.

24. Epstein, "Ruling the World of Money," p. 46.

25. "Dr. Leutwiler's Outburst," *International Currency Review* 15 (September 1983): 45.

Chapter 10

1. M. Mayer, *The Fate of the Dollar* (New York: Times Books, 1980), p. 325.

2. See the Federal Reserve Board, *Federal Reserve Bulletin* (Washington, D.C.: U.S. Government Printing Office, various issues).

3. Data are from the International Monetary Fund, *International Financial Statistics* (Washington, D.C.: IMF, January 1978).

4. Estimate cited in "Diversification Versus Substitution," *Euromoney*, October 1979, p. 10.

5. Reported in ibid., p. 16.

6. "The OPEC Decade," *Economist*, 29 December 1979, p. 45.

7. Federal Reserve Board, *Federal Reserve Bulletin*, January 1982.

8. "The OPEC Decade," p. 45.

9. Paul A. Volcker, Statement before the Committee on Banking, Finance and Urban Affairs, 2 February 1983.

10. *World Financial Markets* (New York: Morgan Guaranty Trust, June 1983).

11. See the International Monetary Fund, *Annual Report* (Washington, D.C.: IMF, 1983).

12. Paul A. Volcker, Statement before the Committee on Banking, Finance and Urban Affairs, 2 February 1983.

13. See "Problem Latin Loans Hit 7% Level," *American Banker*, 5 December 1983, p. 3.

14. *World Financial Markets*.

15. Calculated from U.S. GNP data in Council of Economic Advisors, *Economic Report of the President* (Washington, D.C.: U.S. Government Printing Office, February 1983).

16. "Dealing With Deficits: Learn to Accept Them," *Wall Street Journal*, 14 December 1981, p. 1.

17. *Monetary Trends* (St. Louis: Federal Reserve Bank of St. Louis, September 1982).

18. See "The Crash of 198?," *Economist*, 16 October 1982, p. 23.

19. *Euromoney*, various issues, from January 1980 to December 1982.

Chapter 11

1. "Under Sense of Unease, IMF Wrestles With Quotas and Crisis Fund," *Washington Post*, 6 September 1982, p. A9.

2. "U.S. Is Said to Support IMF Rise," *New York Times*, 4 December 1982, p. D1.

Notes

3. See "Banks Charging $800 Million for Help to Mexico," *Washington Post,* 20 March 1983, p. G1.

4. "The World's Biggest Borrower Hasn't Got a Bean," *Economist,* 21 August 1982, p. 49.

5. "Debtors' Prism," *Economist,* 11 September 1982, p. 13.

6. "Lenders' Jitters," *Wall Street Journal,* 15 September 1982, p. 29.

7. "A Mexican Model for Latin American Debt," *Economist,* 3 September 1983, p. 69.

8. "How Brazil Struggled to Refinance Its Debt but Met Early Failure," *Wall Street Journal,* 29–30 August 1983, p. 1.

9. "How They Tried to Rescue Brazil," *Euromoney,* October 1983, p. 79.

10. See "Hope Stirs for World Debt Relief," *New York Times,* 19 April 1983, p. D1.

11. See "IMF Funds Are Linked to Housing," *Washington Post,* 15 August 1983, p. A1.

12. Eliot Janeway, "The Prince of Pessimism Says the System Could Topple," *Washington Post,* 14 August 1983, p. C1.

13. "Argentina Fires a Shot in the Debtors' Revolt," *Business Week,* 17 October 1983, pp. 47, 51.

14. "Massive IMF Housing Compromise Is Approved By Senate, 67 to 30," *Washington Post,* 18 November 1983, p. D1.

15. "What Bankers Finally Learned from the Debt Crisis," *Business Week,* 5 December 1983, p. 68.

16. William Greider, "Uncle Sucker," *Rolling Stone,* reprinted in *Seattle Times,* 27 November 1983, p. A15.

Chapter 12

1. Charles P. Kindleberger, *Manias, Panics, and Crashes* (New York: Basic Books, 1978).

2. D. H. Meadows, et al., *The Limits to Growth* (New York: Universe Books, 1972).

3. "Why Venezuela Is the New Sick Man of Latin America," *Business Week,* 5 December 1983, p. 98.

4. "A Hot New Market in Swapping High-Risk Debt," *Business Week,* 5 December 1983, p. 144.

5. "Games Bankers Play," *Forbes,* 5 December 1983, p. 186.

6. These and other bank data for September 30, 1983, are from "Problem Loans Hit 7% Level," *American Banker,* 5 December 1983, p. 3.

7. "Euromoney Five Hundred," *Euromoney,* June 1983, p. S145.

8. Data are based on an article by Fred R. Bleakley, "Waiting for the Big Banks to Revive," *New York Times,* 13 November 1983.

9. "Problem Loans Hit 7% Level."

10. See "What Debt Crisis?" *Wall Street Journal,* 20 April 1983, p. 28.

11. See Federal Reserve Board, *Federal Reserve Bulletin* (Washington, D.C.: U.S. Government Printing Office, November 1983), pp. 9–52.

12. John Donne, *Devotions on Emergent Occasions* (see John Bartlett, *Bartlett's Familiar Quotations,* 15th ed., Boston: Little, Brown, 1980).

Index

Index

Index

Index

Index

Index